IV International Conference on Modern Technologies for Non-Destructive Testing 2015

IOP Conference Series: Materials Science and Engineering Volume 132

Tomsk, Russia
5 – 10 October 2015

ISBN: 978-1-5108-2535-2
ISSN: 1757-8981

Printed from e-media with permission by:

Curran Associates, Inc.
57 Morehouse Lane
Red Hook, NY 12571

Some format issues inherent in the e-media version may also appear in this print version.

Printed by Curran Associates, Inc. (2016)

For permission requests, please contact the Institute of Physics at the address below.

Institute of Physics
Dirac House, Temple Back
Bristol BS1 6BE UK

Phone: 44 1 17 929 7481
Fax: 44 1 17 920 0979

techtracking@iop.org

Additional copies of this publication are available from:

Curran Associates, Inc.
57 Morehouse Lane
Red Hook, NY 12571 USA
Phone: 845-758-0400
Fax: 845-758-2633
Email: curran@proceedings.com
Web: www.proceedings.com

TABLE OF CONTENTS

IOP Conf. Series: Materials Science and Engineering **132** (2016) 011001 doi:10.1088/1757-899X/132/1/011001

1. Introduction (about the Conference)

This issue of the journal is dedicated to the scientific and applied research presented at the IV International Conference «Modern Technologies for Non-Destructive Testing». The conference was held on October 5–10, 2015, in Tomsk, Russia.

The conference brought together experts from different countries and organizations who received a great opportunity to share knowledge in sessions, to initiate discussions on topics of interest to all participants.

The aim of the conference was to support young talented people, to involve pupils, students, post-graduate students and young scientists in the solution of actual scientific problems of non-destructive testing, quality management and new educational technologies.

The main goals of the event were exchanging information and discussing topics related to non-destructive testing, quality management, new educational technologies and widespread implementation of the obtained results.

The key topics of the conference were advanced technologies in non-destructive testing, effective quality management systems, new educational technologies in the field of control and management, and management concepts of modern economy.

The Institute of Non-Destructive Testing of the National Research Tomsk Polytechnic University was the Conference Organizer.

IOP Conf. Series: Materials Science and Engineering **132** (2016) 011001 doi:10.1088/1757-899X/132/1/011001

2. **List of participants (with email)**

№		
1.	R. Fediuk, A. Yushin	E-mail: roman44@yandex.ru
2.	S. Osipov, I. Podshivalov, O. Osipov, A. Zhantybaev	E-mail: osip1809@rambler.ru
3.	O. V. Galtseva, S. V. Bordunov, N. M. Natalinova, S. V. Mazikov	E-mail: rabota2013tpu@mail.ru
4.	A. Lider, V. Larionov, M. Kroening, V. Kudiiarov	E-mail: victor31479@mail.ru
5.	L. Kozlova, E. Bolovin, L. Payuk	E-mail: kozlovale@tpu.ru
6.	R. Damineva, R. Daminev, O. Karimov	E-mail: daminevarm@mail.ru
7.	M. Grigoriev, L. Babich	E-mail: mishatpu@mail.ru
8.	S. Chakhlov, P. Anpilogov, A. Batranin, S. Osipov, Sh Zhumabekova, I. Yadrenkin	E-mail: chakhlov@tpu.ru
9.	M. Khasanov, V. Shagapov	E-mail: hasanovmk@mail.ru
10.	I. Matveev, E. Siemens, A. Yurchenko, D. Kuznetsov	E-mail: niipp@inbox.ru
11.	Yu. Britova, V. Dmitriev, T. Kostyuchenko	E-mail: britova@tpu.ru
12.	A. Abashkin, S. Osipov, S. Chakhlov, A Shteyn	E-mail: chakhlov@tpu.ru
13.	Yu. Bezborodov, A. Selsky, N. Lysyannikova, Ye. Kravtsova, A. Lysyannikov, V. Shram, Yu. Kaiser, M. Merko	E-mail: nataly.nm@mail.ru
14.	A. Lysyannikov, P. Kondrashov, P. Pavlova	E-mail: lysyannikov.alek@mail.ru
15.	Ye. Kravtsova, V. Shram, N. Lysyannikova, Yu. Bezborodov, A. Selsky, A. Lysyannikov	E-mail: Rina_986@mail.ru
16.	Yu. Serebrenikova, A. Lysyannikov, Yu. Kaizer, R. Zhelykevich, M. Plakhotnikova, N. Lysyannikova, M. Merko, I. Merko	E-mail: jl.serebro@mail.ru
17.	S. Obukhov, I. Plotnikov, M. Kryuchkova	E-mail: igorplt@tpu.ru
18.	E. Boretsky, M. Egorova	E-mail: eboretsky@mail.ru
19.	E. Yankovich, O. Nevidimova, K. Yankovich	E-mail: yankovich@tpu.ru
20.	N. Kalinichenko, I. Lobanova, A. Kalinichenko, E. Loboda, T. Jakubec	E-mail: konareva_irina_88@mail.ru
21.	V. Shram, Ye. Kravtsova, A. Selsky, Yu. Bezborodov, N. Lysyannikova, A. Lysyannikov	E-mail: shram18rus@mail.ru
22.	V. Sorokin, V. Bespalov	E-mail: sorvb@tpu.ru
23.	I. Plotnikova, L. Redko, O. Titova, O. Chaykovskaya	E-mail: inna@tpu.ru
24.	E. Pomishin, L. Yavorovich	E-mail: lusi@tpu.ru
25.	R. Zhelykevich, A. Lysyannikov, Yu. Kaiser, Yu. Serebrenikova, N. Lysyannikova, V. Shram, Ye. Kravtsova, M. Plakhotnikova	E-mail: lysyannikov.alek@mail.ru
26.	N. Lysyannikova, B. Kovalski, Yu. Bezborodov, A. Lysyannikov, Ye. Kravtsova, V. Shram, M. Kovaleva	E-mail: nataly.nm@mail.ru
27.	Y. Salchak, T. Tverdokhlebova, S. Sharavina, A. Lider	E-mail: yasalchak@gmail.com
28.	A. Sagdatullin, E. Muravyova, M. Sharipov	E-mail: saturn-s5@mail.ru
29.	N. Natalinova, N. Ilina, E. Frantcuzskaia	E-mail: natalinova@tpu.ru

IOP Conf. Series: Materials Science and Engineering **132** (2016) 011001 doi:10.1088/1757-899X/132/1/011001

3. Program Committee

- ***Aleksandr S. Sigov*** - Dr., Prof., Academician, President of Moscow State Institute of Radio-Engineering Electronics and Automation, General Chair, Moscow, Russia

- ***Eduard S. Gorkunov*** - Dr., Prof., Academician, Director of Institute of Engineering Science, UB RAN, Yekaterinburg, Russia

- ***Anatoliy P. Surzikov*** - Dr., Prof., Deputy Director of Institute of Non-destructive testing for Science, Head of Department of Physical methods of Non-destructive testing, Tomsk Polytechnic University, Vice-chairman, Tomsk, Russia

- ***Boris I. Kapranov*** - Dr., Prof., Sector leader Technical Tomography and Introscopy, Lab №40, Tomsk Polytechnic University, vice-chairman, Tomsk, Russia

- ***Vladimir P. Vavilov*** - Dr., Prof., Head of Thermal Control Methods Lab №34 Tomsk Polytechnic university, Vice-chairman, Tomsk, Russia

- ***Eduard Siemens*** - Prof., Anhalt University of Applied Sciences, Bernburg, Germany

- ***V.A. Pomogaev*** - researcher of Radiation Lab, postdoc of Chemical and Biomedical engineering department, Notherdam University, Indiana, USA

- ***Jao Ven*** - Prof., Northeastern University, Massachusetts, USA

- ***Thomas Bohlen*** - Dr., Prof., Karlsruhe Institute of Technology, Karlsruhe, Germany

- ***André Kurzmann*** - Institute of Geophysics, Karlsruhe Institute of Technology, Karlsruhe, German

- ***Klaus Eichhorn*** - researcher, Institute of learning photons and synchrotron emission, Karlsruhe Institute of Technology, Karlsruhe, Germany

- ***Oldrih Stary*** - Faculty of Electrical Engineering, Czech Technical University in Prague, Prague, Czech Republic

- ***S.B. Habibova*** - Chief of International Relations Department, The Ministry of Education of the Republic of Tajikistan, Tajikistan

- ***Ilkay Yilmaz*** - Prof., vice-chairman, Department of Economics, Mersin University, Turkey

- ***B.T. Torobekov*** - PhD, Professor, Vice-rector for academic work Kyrgyz State Technical University after I. Razzakov, Kyrgyzstan

- ***Aleksandr V. Rogachev*** - PDr., Prof., Rector of Francisk Skorina Gomel State University, Gomel, Republic of Belarus

- ***Alexey Elyashevich*** - Dr., Prof., Peter the Great St.Petersburg Polytechnic University, St.Petersburg, Russia

- ***Petr Ya. Kalita*** - DPhD, academician of Ukraine Academy, president of Ukraine quality association, Academy of Quality Problems, Ukraine

IOP Conf. Series: Materials Science and Engineering **132** (2016) 011001 doi:10.1088/1757-899X/132/1/011001

- ***Tatiana V. Fursa*** - Dr., leading researcher Electronics, Dielectrics and Semiconductors Lab Tomsk Polytechnic university, Tomsk, Russia

- ***Olga N. Chaykovskaya*** - Dr., Prof., Head of Faculty of physics, Tomsk State University, Tomsk, Russia

- ***Inna V. Plotnikova*** - PhD, Associate Professor of Department of Physical methods of Non-destructive testing Tomsk Polytechnic University, Tomsk, Russia

- ***Olga V. Galtseva*** - PhD, Associate Professor of Department of Physical methods of Non-destructive testing Tomsk Polytechnic University, Tomsk, Russia

- ***Alexey N. Kalinichenko*** - PhD, Associate Professor of Department of Physical methods of Non-destructive testing Tomsk Polytechnic University, Tomsk, Russia, conference chair

Acknowledgements

The Program Committee thanks all participants, organizers and sponsors of the conference. This conference is not only a scientific event allowing the participants to know up-to-date technologies, new investigations and innovative ideas, but a regular meeting of colleagues, who have become trusted friends, partners and colleagues in a common cause.

IOP Conf. Series: Materials Science and Engineering **132** (2016) 011002 doi:10.1088/1757-899X/132/1/011002

Peer review statement

All papers published in this volume of *IOP Conference Series: Materials Science and Engineering* have been peer reviewed through processes administered by the proceedings Editors. Reviews were conducted by expert referees to the professional and scientific standards expected of a proceedings journal published by IOP Publishing.

IOP Conf. Series: Materials Science and Engineering **132** (2016) 012001 doi:10.1088/1757-899X/132/1/012001

Modern Technologies of Nondestructive Testing of Construction Materials

R Fediuk[1] and A Yushin[2]

[1]Senior Lecturer, Engineering School, Far Eastern Federal University, Vladivostok, Russia
[2]Student,Engineering School, Far Eastern Federal University, Vladivostok, Russia

E-mail: roman44@yandex.ru

Abstract. The article presents the modern methods of research of building materials (such as styrofoam, cement, concrete admixtures, etc.), applied in the Far Eastern Federal University. The latest equipment described for these studies and modern methods of testing.

1. Introduction

In recent years, to determine the processes of formation structure in products from inorganic building materials binders are widely used non-destructive inspection methods. These methods best meet the requirements of continuous monitoring of the properties of construction materials, the study of their physical and chemical properties. This test is not accompanied by the destruction of the structure of the material and do not require bookmarks foreign bodies in the sample, as well as allow individuals to obtain objective indicators, are valid not only for small but also for large samples.

2. Basic information

2.1. Chromatography of destruction polystyrene

To assess the quantitative and qualitative composition of toxic substances released during of destruction polystyrene used PGC-1 – autonomous wearable gas chromatograph using a highly sensitive photoionization detector (PID) (figure 1)..

Figure 1. Chromatograph PGC – 1.

For more accurate results, samples were also carried out research on gas chromatography-mass spectrometer Shimatsu QP – 5050A (figure 2).

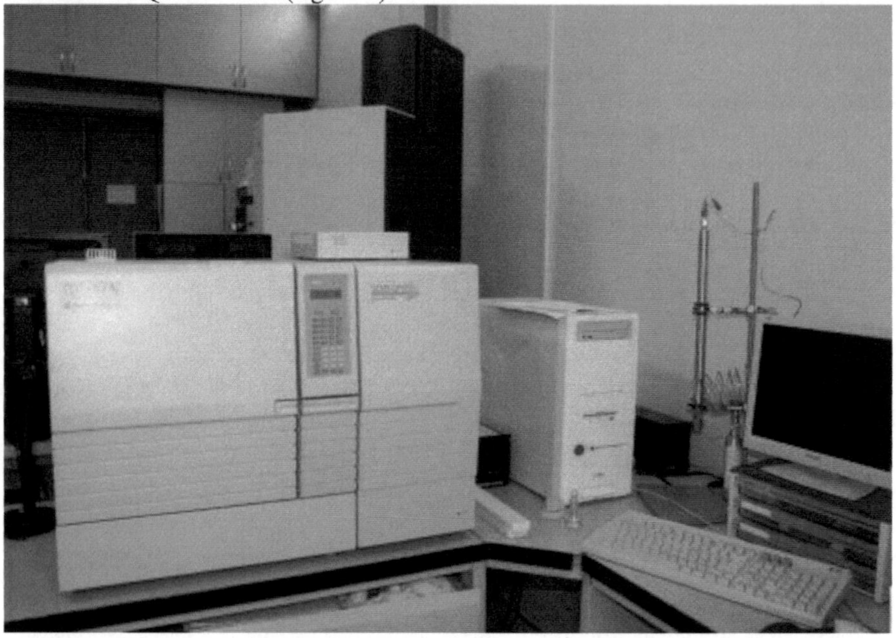

Figure 2. Gas chromatography – mass spectrometer Shimatsu QP – 5050A.

Flexible quadrupole gas chromatography-mass spectrometer along with electron impact ionization (EI) and the system of positive chemical ionization (PCI) for precise determination of molecular weight also has a system of negative chemical ionization (NCI) for highly sensitive analysis of electrophiles. It has a computer - optimizable ion source and the system of magnetic lenses. Quick and clean (150 liter/sec) turbomolecular vacuum pump provides for chemical ionization and ability to work with packed columns within 5 minutes after the activation. Mass Range – 900 Daltons ($1.5 \cdot 10^{-21}$ g) is achieved by means of a detector with low noise. There is a possibility of direct sample introduction, the use of capillary and packed columns. There is a quick switch from automatic mode negative chemical ionization (NCI) in the electron impact ionization (EI) without replacing the components of the ion source. There is a direct connection Purge-and-Trap (VOC concentrator column) without introducing a jet separator. The scanning speed is 6750AEM / s or 50 scans / s, the concentration dynamic range – 106.

2.2. The study of morphological features of the microstructure using scanning electron microscopy
The microstructure of samples was performed using a scanning electron microscope Carl Zeiss CrossBeam 1540XB (figure 3).

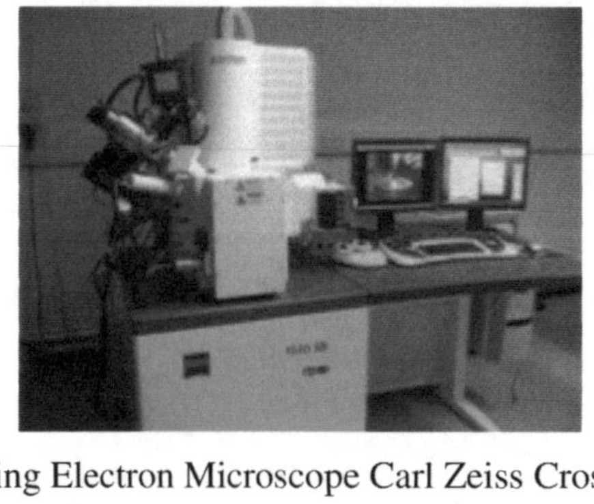

Figure 3. Scanning Electron Microscope Carl Zeiss CrossBeam 1540XB.

IOP Conf. Series: Materials Science and Engineering **132** (2016) 012001 doi:10.1088/1757-899X/132/1/012001

Microstructural studies and quantitative analysis of the microstructure of the samples were carried out using the hardware-software complex, which includes a laser diffraction particle size analyzer "Analysette 22" model NanoTech (Germany) [1].

For the analysis was used specially developed software package automated processing, enabling you to get almost all the morphological parameters of the microstructure (the size and shape of the structural elements, their orientation in space), and to assess the value of porosity and specific surface area.

2.3. X-ray analysis

Study of the qualitative and quantitative composition and properties of the starting materials and composite binder was performed using standard methods. Investigation of the mineral composition and the structure is performed using X-ray diffraction on the powder X-ray diffractometer D8 Advance Company Bruker AXS (figure 4).

Figure 4. X-ray powder diffractometer D8 Advance.

X-ray diffraction - a non-destructive method for analyzing substances in powder form. It provides the ability to conduct qualitative and quantitative analysis of crystalline phases of establishing the crystal structure of inorganic, organic, organic and metal complex compounds in polycrystalline form, establishing the composition of polycrystalline materials, the degree of crystallinity of polymers [2].

In the study used the database ICDD PDF–2 and Crystallography Open Database. Quantitative phase analysis is possible for 345 industrially important phases contained in the database TOPAS Structure Database, as well as the phase of known structure [3].

2.4. Differential thermal analysis and thermogravimetry

Depending on the latched indicators when heated distinguish the following methods of thermal analysis:

- Differential thermal analysis (DTA) - shows the change in the energy system (sample);
- Thermogravimetry (TG) - change in mass [4].

Derivatograms samples were obtained on a thermogravimetric analyzer, Shimadzu DTG–60H (figure 5). Program heating furnaces from 20 to 1500°C is carried out by electronically heater at 20 °C / min. Platinum thermocouple with an accuracy of 5 ° C temperature measurement (T), the scan speed at 2.5 mm / min a four-channel recorder signal is recorded on the paper. The temperature difference (ΔT) between the studied substance and a standard proportional to the thermal effect is recorded in the form of DTA curve (sensitivity of 0.5 mV). Simultaneously with the DTA curve is being recorded curve of weight loss (TG) and its derivative (DTG) (sensitivity of 0.5 mV), 113 mg sample weight. Weighing accuracy of 0.05 mg [5].

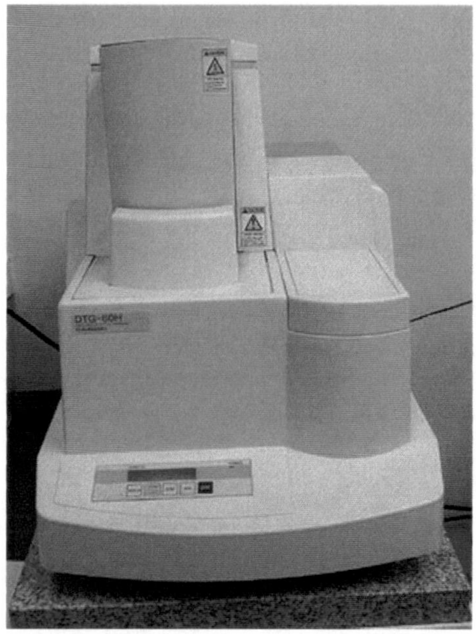

Figure 5. Thermogravimetric Analyzer Shimadzu DTG-60H.

2.5. Determination of radioactive materials

In the literature review, it was found that materials such as crushed granite screenings and ash TPP have some radioactivity. Due to the fact that the concept of the modern building materials science is aimed at creating conditions for the environmental safety of the home, it was necessary to quantitatively check the background radiation of these materials [6, 7].

The specific (volumetric) activity of beta- and gamma-emitting nuclides in the samples was determined by counting spectrometric method using a universal spectrometer complex USC "Gamma Plus" (figure 6).

Figure 6. Universal spectrometer complex USC "Gamma Plus".

The study was conducted according to the requirements of the following standard documents: GOST 27451-87 "Means of measuring ionizing radiation. General technical conditions", GOST 26864–86 "Spectrometers of energies of ionizing radiation. Methods of measurement of basic parameters", TU 4362-002-46554900-06 (PLUS 412131.002TU) "complex universal spectrometry USC" Gamma Plus. "Specifications".

The principle of operation is based on the complex transformation in the working volume of the detector of gamma rays and beta particles in the flashes of light (scintillation), the intensity of which is proportional to the energy lost by the gamma-quantum or beta-particle detector.

Lights flash, falling into a photoelectron amplifier tube (PMT), converted into a stream of electrons, which multiply under the influence of the applied potential difference, resulting in at the output of the PMT generated pulses of electric current, whose amplitude is proportional to the particle energy lost in the detector. This fact provides the fundamental ability to measure the energy spectrum of the detected gamma - or beta radiation.

2.6. Determination of particle size materials

Particle size was determined by laser particle size distribution of powder materials, allows you to determine the particle size and the percentage of their content in the analyzed material.

For these purposes, used laser diffraction particle size analyzer "Analysette 22" model NanoTech (Germany) (figure 7).

Figure 7. Laser diffraction particle analyzer "Analysette 22" NanoTec.

Analyzer for determining the particle size distribution by laser diffraction employs the physical principle of scattering of electromagnetic waves. Particles in a parallel laser beam scatter light at a constant solid angle, which depends on the particle diameter. The lens collects the scattered light annularly detector, which is mounted in the focal plane of the lens. Non-scattering light always converges at the focal point on the optical axis.

With complex mathematics of the intensity distribution of the scattered light can calculate the particle size distribution of the scattering particles in the team. The result is a laser diffraction particle diameter, the diameter of which is equivalent to a sphere with the same distribution of scattered light. Measured volume mean diameter and the resulting particle size distribution, this is distribution volume.

2.7. Determination of the specific surface

Specific surface area binders and mineral additives used were measured according to GOST 310.2-81, using an instrument PSC-11, which operates on the principle of air permeability through the layer previously compacted material (figure 8).

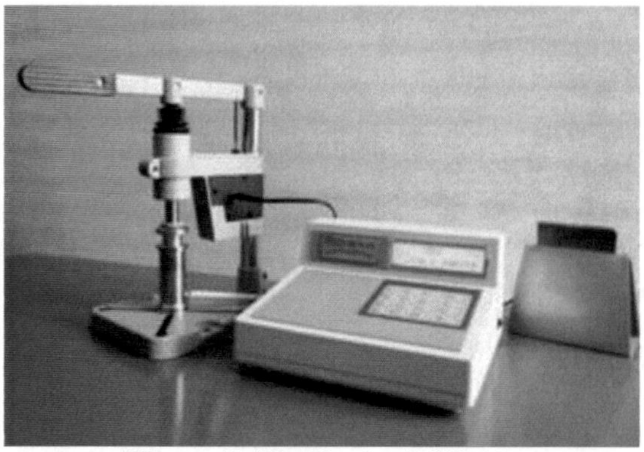

Figure 8. The device PSC-11.

All measurements and calculations in the PSC-11 automated, eliminating the factor of human error. The measurement results are displayed on the display. The unit of analysis of variance PSC-11 is used internationally accepted method gas permeability Kozeny and Carman. Coefficient of permeability of samples with open porosity or powder layer is determined by the duration of the filtration air therethrough at a fixed initial and final vacuum in the working volume of the device.

The test particulate material is introduced into the sleeve layer of a certain thickness, after a vacuum is created by means of a water jet pump. Using the regulator device when the cock is open, a vacuum is created in which the fluid in the closed knee gauge aspirator begins to move from the closed knee open due to the level difference, thus there is the passage of air through a bed located in the liner material. The duration of the passage of a certain volume of air through a bed of particulate material is dependent on the value of its specific surface area, which allows to calculate its value.

The calculation is made using equation 1:

$$S = K \frac{M\sqrt{t}}{m} \tag{1}$$

where S – the value of specific surface area, K – constant of the device (the device is indicated in the passport); M – value depending on the height of the powder layer and the ambient temperature (determined by the table); t – time of passage of fluid between the risks; m – mass of sample [g].

3. Summary

According to the results of series of experiments the amendment of the well-known Fuchsian equation is calculated for calculation of the resistance of fibrous air filter with regard to form and surface defects of the fibers.

References

[1] Boukherroub N *et al* 2012 *EPJ Web of Conferences* **29** 00010
[2] Kezrane M *et al* 2012 *Journal of Alloys and Compound* **536** 304–307
[3] Predescu C *et al* 2012 *Metalurgia International* **18** 65–68
[4] Akkouche K *et al* 2011 *Journal of Magnetism and Magnetic Materials* **323** 2542–2548
[5] Nicolicescu C *et al* 2011 *Materials Science Forum* **672** 219–222
[6] Nicolicescu C *et al* 2010 *Optoelectronics and Advanced Materials, Rapid Communications* **4** 1408–1414
[7] Hales S J *and* Vasquez P 2003 *Gamma Titanium Aluminides* 305–310

IOP Conf. Series: Materials Science and Engineering **132** (2016) 012002 doi:10.1088/1757-899X/132/1/012002

Informativeness Improvement of Hardness Test Methods for Metal Product Assessment

S Osipov[1], I Podshivalov[2], O Osipov[3] and A Zhantybaev[4]

[1]Leading Researcher, National Research Tomsk Polytechnic University, Tomsk, Russia
[2]Associate Professor, Tomsk State University of Architecture and Building, Tomsk, Russia
[3]Engineer, National Research Tomsk Polytechnic University, Tomsk, Russia
[4]Master Student, National Research Tomsk Polytechnic University, Tomsk, Russia

E-mail: osip1809@rambler.ru

Abstract. The paper presents a combination of theoretical suggestions, results, and observations allowing to improve the informativeness of hardness testing process in solving problems of metal product assessment while in operation. The hardness value of metal surface obtained by a single measurement is considered to be random. Various measures of location and scattering of the random variable were experimentally estimated for a number of test samples using the correlation analysis, and their close interaction was studied. It was stated that in metal assessment, the main informative characteristics of hardness testing process are its average value and mean-square deviation for measures of location and scattering, respectively.

1. Introduction

The assessment of metal products while in operation shows their degradation due to such negative factors [1–3] as wear abrasion, corrosions, and fatigue. One of the metal assessment methods used for products while in work and operation is their hardness testing [5–6]. For the last two decades, metal assessment methods have been intensively developed as well as instruments and scopes of their application [7–9]. The traditional use of durometers implies measurement of product at its several points, and the average value of obtained results should be used to assess the metal hardness. Ekici, Starikov, Muzyka, *et al.* [10–12] showed that the use of the additional informative parameter identified during the hardness testing process, namely, the hardness scattering, allows the improvement of the metal assessment method. Until recently, the implementation of this technique has been restricted by the instrument unavailability. However, the production of high-end portable durometers with a lower effect on test samples eliminated this restriction, and a parallel evaluation of the average value of hardness and its certain measure of scattering is now being used in transportation, construction and other industries [13–15]. The literature review concerning the use of other informative hardness parameters showed the insufficiency of the data.

2. Theory

Mechanical stresses, corrosions, and intensive tensile and compressive forces result in structural modifications of the surface layers of metals and alloys.

The following variants are the most visible manifestation of these changes: • surface roughness (the kind, direction, and level of mechanical stresses as well as their duration define the degree of the surface modification. It should be noted that surface roughness can be both increased and decreased); • small disoriented fatigue cracks; • explicit, oriented and rather deep cracks. Such cracks are usually induced by metal-to-metal contact of parts by repeated trajectories; • corrosion spots (oxides, salts of mother metal(s)). The specific density (relative fraction of corrosion spot area) and the size of spots depend on the duration the metal product exposure to such hazardous factors as humidity, acid and acid solution vapors, alkalis, and salt solutions; • oxide, salt and other films produced by corrosion.

The value of hardness H obtained by a single measurement is naturally considered to be random. The random variable x is completely defined in case its probability density function $F(x)$ is known [16]. The random variable collection of characters is divided into two large groups, namely: measure of location and measure of scattering. In the work of Korn *et al.* [16], measures of location of random variable x include the average value $\bar{x}$, median Mex and mode Mox. Measures of scattering of random variable x include dispersion $\sigma^2 x$, mean-square deviation σx, coefficient of variation Vx, coefficient of skewness Ax, and coefficient of excess Ex.

Note 1 In scientific literature, the distribution function of the random variable H is described by different laws such as lognormal, and Weibull and Gumbel distributions. These laws are two-parameter distributions which comprise a measure of location and a measure of scattering.

All mentioned measures of location and scattering including those in Note 1 can be used for metal assessment when applied to the random variable H.

Note 2. The degree of metal surface uniformity of hardness, sometimes called the coefficient of uniformity k [12–14], is connected with values of measure of location.

Note 3. The estimation of hardness value obtained by a single measurement depends on the contact area between the indenter and the material being tested.

From Note 3 it follows that the smaller the size of indenter the more evident is the scattering of hardness. Presently, the size of indenters and their penetration force can be varied. Depending on the size of indentation, hardness, micro-, and nano-hardness can be measured.

Let us put forward the basic statements which allow a substantiation of character of the random variable H used in the capacity of metal product assessment provided by hardness test methods.

Statement 1. The change of metal surface properties under any impact conditions is caused by its structural modification.

Statement 2. For each of variants of visible manifestation of changes stated above, the surface layer modified during the product operation will possess at least one hardness parameter regarded as a random variable, dissimilar to the respective one of the original metal.

Note 4. One informative parameter of hardness H may turn to be insufficient to provide the metal assessment. In this case, it is advisable to use two or more informative parameters.

Statement 3. A sampling parameter of the random variable H selected from the collection of those to be analyzed can be informatively excessive in case it is closely connected with one or more residual parameters.

Close connection between random variables x and y is evaluated using the correlation factor r_{xy}. Parameters x and y are considered to be closely connected in case the correlation factor r_{xy} satisfies the condition of $r_0 \leq |r_{xy}| \leq 1$. The lower bound r_0 of this interval is determined experimentally. Since the authors present technical measurements, the value r_0 ranges between 0.85–0.95.

It is logically to select such a parameter which is easy to identify for further analysis or some arguments can be put forward in favor of its application in case it possesses an explicit physical interpretation.

Consequence of Statement 3 In metal assessment, the sampling y parameter of the random variable H is additional to the informative parameter x in case it is independent of x parameter.

Let us assume that connection between x and y parameters are negligible at the fulfillment of $0 \leq |r_{xy}| \leq r_1$ condition. The lower bound r_1 of this interval is also determined experimentally. The value r_1 ranges between 0.1–0.2.

IOP Conf. Series: Materials Science and Engineering **132** (2016) 012002 doi:10.1088/1757-899X/132/1/012002

3. Experimental

With a view to assess the informativeness of parameters describing the random variable H, a series of hardness tests was conducted for steel specimens exposed to different types of external actions.

Specimens of different steel types were tested having average value of hardness ranging from 80 to 200 HB. Calibration measurements were conducted by Brinell hardness of 100 and 184 HB. Hardness measurements are characterized by the high-quality treatment of the surface. Therefore, the effect exerted by the surface roughness on the hardness properties assessment as a random variable, is minimized. Hardness tests were carried out with the dynamic indenter TEMP-4 in compliance with recommendations given by technical specifications and guidelines.

The original sampling was provided for each test specimen using Brinell hardness $(H_1, H_2, H_3, ..., H_n)$, where n is the volume of sampling ($n \geq 200$).

In all, seven specimens were tested. Measures of location and scattering of the random variable H were estimated for each specimen. All sampling parameters were subsequently considered as random variables. In accordance with the Consequence of Statement 3, the independence (level of dependence) of these values was verified by couples of parameters.

At the first stage of the experimental data processing the following parameters of the random variable were estimated: the average value $\overline{H}$; median MeH; and mode MoH. These parameters can be obtained from [16]:

$$\overline{H} = \frac{\sum_{i=1}^{n} H_i}{n}, \quad \sum_{H_i \leq \mathrm{Me}H} 1 = \sum_{H_i \geq \mathrm{Me}H} 1, \quad f(\mathrm{Mo}H) = \max f(H) \quad . \tag{1}$$

Along with measures of location, measures of scattering were determined: mean-square deviation σ_H, coefficient of variation V_H, coefficient of skewness A_H, and coefficient of excess E_H. Scattering parameters [16] can be obtained form

$$\sigma_H = \sqrt{\frac{\sum_{i=1}^{n} \left(H_i - \overline{H}\right)^2}{n}}, \quad V_H = \frac{\sigma_H}{\overline{H}}, \quad A_H = \frac{m_3}{m_2^{3/2}}, \quad E_H = \frac{m_4}{m_2^2} - 3. \tag{2}$$

where m_k is the central k-order sampling moment of the value analyzed. It can be found from [16]:

$$m_k = \frac{\sum_{i=1}^{n} \left(H_i - \overline{H}\right)^k}{n} \quad . \tag{3}$$

According to note 2, parameters of the assumed H random distribution can act as measures of location and scattering. In the number of scientific papers devoted to the study of strength of materials, it is assumed that H or ln$H$ random variables are Weibull-distributed.

The random variable x is Weibull-distributed [16] in case its distribution function is described by equation 4:

$$F(x) = 1 - \mathrm{e}^{-\left(\frac{x}{\lambda}\right)^k} \quad . \tag{4}$$

In addition to (1) and (2), the location of the random variable H will be characterized by λ_H and $\lambda_{\ln H}$ parameters, while its scattering by k_H and $k_{\ln H}$ parameters.

Weibull distribution parameters k and λ can be obtained using the method of matching moments. Two initial moments $\overline{x}$ and $\overline{x^2}$ are to be obtained for the random variable x. Weibull distribution parameters k and λ are obtained from [17]:

$$\frac{\overline{x^2}}{\left(\overline{x}\right)^2} = \frac{\Gamma\left(1+\frac{2}{k}\right)}{\Gamma^2\left(1+\frac{1}{k}\right)}, \quad \lambda = \frac{\overline{x}}{\Gamma\left(1+\frac{1}{k}\right)}, \quad \Gamma(z) = \int_0^\infty t^{z-1}e^{-t}dt. \tag{5}$$

In [11–13], the authors describe the method of estimation of the coefficient $k_{\ln H}$:

$$\overline{\ln H} = \frac{\sum_{i=1}^{n} \ln H_i}{n}, \quad \sigma_{\ln H} = \sqrt{\frac{\sum_{i=1}^{n} \left(\ln H_i - \overline{\ln H}\right)^2}{n-1}}. \tag{6}$$

Gumbel approximation for k parameter is as follows:

$$k = \frac{0{,}4343 d_n}{\sigma_{\ln H}}. \tag{7}$$

It is assumed that the larger values of k_H and $k_{\ln H}$ coefficients provide the low level of hardness scattering and the better structural organization and the lower degree of damage. Unlike the larger, the smaller values of k_H and $k_{\ln H}$ coefficients provide the highest degree of damage. It was shown in [11–13] that the level of scattering measures of metal property to be identified, including hardness, can correspond to another statistical criteria. In the present work, the authors verify the validity of this statement.

At the second stage, verification of the availability, direction, and closeness of correlation between parameters of the random variable H was carried out using the correlation analysis. In case of sampling $(x_1, x_2,..., x_n)$ and $(y_1, y_2,..., y_n)$ for x and y random variables, the correlation factor r_{xy} takes the form

$$r_{xy} = \frac{\overline{xy} - \overline{x}\,\overline{y}}{\sigma_x \sigma_y}. \tag{8}$$

The sign and the absolute value of a correlation factor describe the direction and the magnitude of the relationship between two variables.

The random variables x and y run over all possible location and scattering parameters, respectively. In equations 2–3, the number n equals to the number of specimens. Seven specimens (n = 7) are sufficient merely for a preliminary study of the problem.

The estimation of the coefficient of uniformity k based on equation 5, is rather complicated and requires a preliminary construction of the additional function of k. In equation 5 this function is expressed by the right hand side of the equation. Mathcad intended for the verification, validation, documentation of engineering calculations, can be an alternative for solving this problem.

4. Results and discussion

Table 1 gives the results of statistical processing of experimental data on Brinell hardness H of specimens.

Table 1 contains the values of parameters of the random variable H location, namely: MeH, MoH, λ_H and $\lambda_{\ln H}$. Also there are given parameters of σ_H, V_H, A, E, k_H, $k_{\ln H}$, which determine the hardness scattering. Parameters of location and scattering of H were identified using equations 1–2 and 5–7.

Table 1. A summary table of sampling parameters of H.

Measure	Parameter	Specimen number						
		1	2	3	4	5	6	7
Location	$\overline{H}$	100.8	184.2	119.2	126.0	126.4	177.8	192.3
	MeH	101.2	184.3	121.7	124.3	128.9	177.6	191.4
	MoH	97.1	184.3	121.7	122.7	134.0	183.2	187.3
	λ_H	111.7	204	132	139.6	139.9	197	213
	$\lambda_{\ln H}$	106.9	189.2	129.3	138.2	139.7	191.5	203.1
Scattering	σ_H	7.13	5.19	12.00	15.93	19.07	15.90	12.06
	V_H	0.0708	0.0282	0.1006	0.1263	0.1509	0.0894	0.0627
	A_H	0.984	0.462	−0.603	0.144	0.839	−0.288	0.197
	E_H	3.90	1.78	0.47	−0.81	2.09	1.37	−0.36
	k_H	17.485	44.917	12.103	9.521	7.935	13.697	19.817
	$k_{\ln H}$	5.104	20.131	5.359	4.446	3.841	6.11	9.014

In order to analyze informativeness of one or another parameter of the random variable HB and identify the excessive parameters, one can use the correlation analysis. It should be noted that the sampling volume is insignificant. However, it may turn to be enough to determine informativeness of the suggested process of hardness testing. The correlation factor r_{xy} is calculated from equation 8. Here random variables x and y run over all parameters of the random variable H, namely: $\overline{H}$, MeH, MoH, λ_H, $\lambda_{\ln H}$, σ_H, V_H, A_H, E_H, k_H, and $k_{\ln H}$. Correlation factor values for hardness parameters are given in table 2.

Table 2. Correlation factor values for hardness parameters.

Parameters	$\overline{H}$	MeH	MoH	λ_H	$\lambda_{\ln H}$	σ_H	V_H	A_H	E_H	K_H	$K_{\ln H}$
$\overline{H}$		1.00	0.99	1.00	1.00	−0.12	−0.55	−0.25	−0.35	0.55	0.64
MeH	1.00		0.99	1.00	1.00	−0.12	−0.54	−0.25	−0.34	0.55	0.64
MoH	0.99	0.99		0.99	0.99	−0.05	−0.48	−0.27	−0.32	0.51	0.61
λ_H	1.00	1.00	0.99		1.00	−0.12	−0.55	−0.25	−0.35	0.55	0.64
$\lambda_{\ln H}$	1.00	1.00	0.99	1.00		−0.03	−0.47	−0.28	−0.39	0.47	0.57
σ_H	−0.12	−0.12	−0.05	−0.12	−0.03		0.88	−0.20	−0.38	−0.80	−0.69
V_H	−0.55	−0.54	−0.48	−0.55	−0.47	0.88		−0.02	−0.18	−0.87	−0.80
A_H	−0.25	−0.25	−0.27	−0.25	−0.28	−0.20	−0.02		0.63	0.17	0.08
E_H	−0.35	−0.34	−0.32	−0.35	−0.39	−0.38	−0.18	0.63		0.19	0.05
k_H	0.55	0.55	0.51	0.55	0.47	−0.80	−0.87	0.17	0.19		0.98
$k_{\ln H}$	0.64	0.64	0.61	0.64	0.57	−0.69	−0.80	0.08	0.05	0.98	

From table 2, the following intermediate conclusions can be drawn relative to mutual parameter dependencies which characterize the random variable H.

1. All parameters of the random variable location, namely $\overline{H}$, MeH, MoH, λ_H, $\lambda_{\ln H}$, are strictly dependent on each other; correlation factors are close to zero; the dependencies are strong and direct.

2. The mean-square deviation σ_H does not depend on any location parameter of the random variable H; the absolute values of the correlation factor do not exceed 0.12.

3. The correlation factor V_H weakly depends on all parameters of location. These dependencies are inverse; the absolute values of the correlation factor range from 0.47 to 0.55.

4. Dependence between skewness A_H and excess E_H coefficients and location parameters is practically absent; the absolute values of the correlation factor are less than 0.39.

5. Dependence between coefficients k_H and $k_{\ln H}$ is direct and takes the interfacial position between medium and weak dependencies; the absolute values of the correlation factor range from 0.57 to 0.64.

6. The scattering parameters show the following dependencies: direct $\sigma_H - V_H$ dependence with 0.88 correlation factor; inverse $\sigma_H - k_H$ dependence with −0.8 correlation factor; V_H dependence on both k_H and $k_{\ln H}$ with –0.8–0.87 correlation factor.

5. Implications

From the intermediate conclusions stated above it follows that:

1. All of $\overline{H}$, MeH, MoH, λ_H, $\lambda_{\ln H}$ parameters can be used as a measure of location of the random variable H. However, it is more feasible to use $\overline{H}$ since it is easily identified. All other location parameters of the random variable can be used to exclude random errors in calculations.

2. Parameters σ_H, A_H, E_H can be used as a measure of scattering. All other scattering parameters show a strong dependence on location parameters. It is more feasible to use a mean-square deviation as a parameter least dependent on any location parameter.

6. Conclusion

The experiments showed that in metal assessment, the main informative characteristics of hardness testing process are its average value and mean-square deviation for measure of location and measure of scattering, respectively. All other parameters can be used to improve the metal assessment methods. For this improvement it is expedient to develop multi-channel durometers, the appropriate algorithms of experimental data visualization and processing.

Acknowledgement

This work was financially supported by The Ministry of Education and Science of the Russian Federation in part of the science program.

References

[1] Jacobs J J *et al* 2009 *J. Am. Acad. Orthop. Surg.* **17 (2)** 69–76
[2] Zhang S W *et al* 2004 *Wear* **256** 226–232
[3] Abosrra L *et al* 2011 *Construction and Building Materials* **25 (10)** 3915–3925
[4] Hashemi S 2011 *Mater. Sci. Eng. A.* **528** 1648–1655
[5] Aboutalebi F H *et al* 2011 *Int. J. Adv. Manuf. Technol.* **53** 157–165
[6] Hołowaty J M *and* Wichtowski B 2013 *Structural Engineering International* **23** 512–518
[7] Parasız S A *et al* 2011 *Journal of Manufacturing Processes* **13** 153–159
[8] Zhang S *et al* 2012 *Machining Science and Technology* **16 (3)** 473–486
[9] Hussainova I *et al* 2011 *International Journal of Materials and Product Technology* **40 (40)** 58–74
[10] Ekici R *et al* 2010 *Materials & Design* **31 (6)** 2818–2833
[11] Starikov M *et al* 2011 *Transport* **26 (3)** 255–262
[12] Muzyka N R *and* Shvets V P 2014 *Strength of Materials* **46 (1)** 105–109
[13] Lebedev A A *et al* 2007 *Strength of Materials* **39 (6)** 567–571
[14] Lebedev A A *et al* 2011 *Strength of Materials* **43 (5)** 506–514
[15] Han W *et al* 2012 *Materials Transaction* **53 (2)** 390–394
[16] Korn G A *and* Korn T M 1994 *Handbook of Mathematics for Engineers and Researchers* (Moscow: Nauka)
[17] Gupta R D *and* Kundu D 2001 *Biometrical Journal* **43 (1)** 117–130

IOP Conf. Series: Materials Science and Engineering **132** (2016) 012003 doi:10.1088/1757-899X/132/1/012003

Improvement of the Quality of Water Purification from Hydrocarbons Using the Fibers from Recycled Thermoplastics

O V Galtseva[1], S V Bordunov[2], N M Natalinova[1] and S V Mazikov[3]

[1]Associate Professor, National Research Tomsk Polytechnic University, Tomsk, Russia
[2]Superviser, Public Corporation Scientific - Innovation Enterprise "Echteh", Tomsk, Russia
[3]Student, National Research Tomsk Polytechnic University, Tomsk, Russia

E-mail: rabota2013tpu@mail.ru

Abstract. Adsorption properties of the polymer fibers are studied. It is shown that polypropylene fiber can be successfully applied for oil spill response for filtration purification of water from hydrocarbons. Polypropylene fibers from waste polymers have higher characteristics of adsorption capacity and degree of purification of water than commercially available fiber sipron.

1. Introduction

At the present time almost all of the extracted oil passes from production sites to consumers through a developed system of field and main pipelines, which because of this are the zone of high ecological danger. Since 100% reliability of any functioning system is almost unattainable, the possibility of an emergency with appropriate consequences always exists. An actual problem for the whole world is fight against spills of oil and oil products at accidents. At oil spill the main danger should be considered the spread of oil on big squares on a surface of the soil and water leading to violations of ecological balance of territories and doing impossible normal functioning of biological systems for a long time. The spillage of oil and oil products in the aquatic environment is extremely dangerous because it may be accompanied by spreading of them over a distance of hundreds of kilometers from the scene of the accident.

Number of consecutive technological operations is applied upon liquidation of oil spills on the water surface: localization of the oil slick by floating booms, pumping of main mass of oil by the high-performance mechanical or vacuum skimmers, equipped by separators for separation of oil from water and use of oil for designated purpose. A layer of oil with thickness up to 1 – 2 mm remains on the water surface as a result of this operation. The final purification of the water surface is carried out by different dispersed oil-absorbing materials (OAM). The first task is usually successfully is solved, but daunting task is the collecting of thin oil films (especially, iridescent films) with thickness comparable with wavelength of the visible spectrum of daylight. The use of skimmers with external source of energy is difficult and often impossible, at the oil spills in wetlands and on areas with difficult relief.

Therefore there are two possible ways: microbiological recultivation and collecting of oil by oil-absorbing materials (sorbents).

Paper describes [1–4] the experience of application of various sorbents which used to collect of spilled oil in the pipe transport. Fibrous materials from thermoplastic materials are the most preferred with the ratio "price of adsorbent/weight of collected oil". Due to their high strength, resistance to high temperatures and corrosive influences the fibrous polymeric fibers are increasingly replacing materials from natural fibres in the industry. The technology for producing fibers from thermoplastic processing waste was developed and it is cheaper by 1.5 – 2 times than the fibers received from primary raw materials [5, 6]. These fibers can be used in the processes of separating of the methane-butane fraction from the natural gas before transport [6] and as oil-absorbing materials at liquidation of emergency floods of oil and oil products and in processes of filtration purification of oil- containing water.

2. Experimental

Further the main characteristics of the fibers obtained by the proposed technology are provided [5, 6]. Fibers based on polypropylene (PP) are resistant in mineral and organic acids, alkalis and they have an upper working temperature about 95 ^{0}C. Fibers of polyethylene terephthalate (PE TP) with an operating temperature of 130-150 ^{0}C are exposed to alkalis. Decrease of temperature border of use of received fibers is caused by partial destruction of polymers at their repeated thermal processing and decrease of their molecular weight. According to the derivatographic analysis the losses of mass of commodity granulated polypropylene are noted at 255 ^{0}C, and loss of weight of 0.3% is recorded at 200 ^{0}C for fiber from secondary polypropylene of the same brand. Commodity PE TP begins to lose weight at 350 ^{0}C, and the fiber obtained from fleck of PE TP (bottles), loses 0.5% of mass already at 250 ^{0}C.

Table 1 shows the modes for producing of fibers from recycled thermoplastics. Research of possibility of recycling PP, PE, polystyrene (PS), PE TP and polyvinyl chloride (PVC) into fibrous materials by technology [5] shows the following.

Table 1. The recommended temperature of recycling thermoplastics into fibers by technology [5, 6].

Initial raw material	Processing temperature, ^{0}C	Diameter of resulting fibers (depending on the modes of processing temperature and rotation speed of reactor), μm
PP of any brands	220–320	0.1–200
PE of any brands	280–320	0.1–250
PS of any brands	180–190	1.0–300
PE TP – bottles	280–290	0.1–300
PP + PS (5-30%)	280–290	20–400

PP and PE TP are processed into the fiber more easily and practically without losses. The strength of fibers made from individual PP and PE TP is sufficient for use as filter nozzles in filters for water and air purification and for the production of nonwovens. The optimum processing temperature of PP is in the range 250-310 ^{0}C. In the temperature range 160 - 250 ^{0}C melt viscosity of PP is high and it does not allow to receive the fiber even at the speed of rotation of reactor of 1400 rpm. The melt of PE TP has a lower viscosity than the melt of PP, therefore, it is possible to produce fibers at 275-290 ^{0}C, i.e. it is at the temperature which higher on 5-15 ^{0}C than its melting temperature. Inevitable small additions of PE in the form of stoppers and rims from the stoppers on the bottles from PE TP almost don't influence on the temperature mode of processing of recycled PE TP into the fiber at the recycling of bottles. It has a positive effect on the mechanical characteristics of the fibers of PE TP because they carry out function of a plasticizer. Additives of secondary PE (up to 30%) into secondary PP behave similarly without influencing on temperature mode of processing of PP into fiber.

Figure 1 shows photos of the fibers obtained in [7, 8]. Investigation of the surface of fibers on the electron microscope at a magnification of 1000 times in the Nano-Center of TPU (figure 1) show that only fibers with diameter less than 30 μm have in the cross section of the circle and smooth the surface. More complicated geometry of fibers in the cross section is observed with the increase of the diameter of the resulting fibers. For example, the shape of tape twisted from two parties on width is associated with lateral pressure on the melt in spinnerets which are directed oppositely to the rotation and non-uniform expansion of a jet of the melt outside the reactor. In addition, cracks and cavities appear on the surface of fibers with a diameter more than 100 μm, which explains the decrease in their strength.

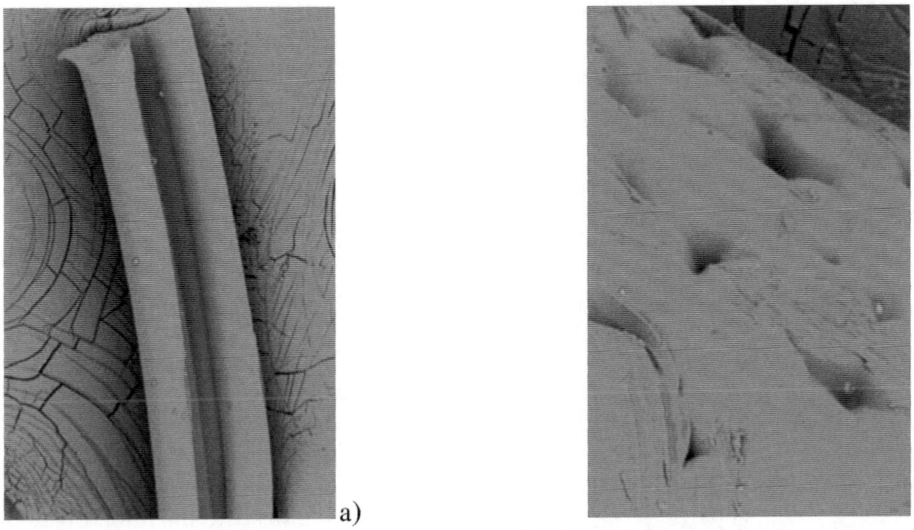

a) b)

Figure 1. Photos of samples of fibers obtained on installation [8], equipped with reactor with r = 125 mm and 156 dies: a – diameter is 100 μm, magnification is 200 times, b – diameter is 100 μm, magnification is 1000 times, cavities are 1–5 μm.

The wettability of the material by the adsorbed liquid determines the value of the regional angle of wetting. It is one of the main characteristics of adsorbent determining its absorbing ability by this liquid. Besides, adsorbents must have the low cost, availability and rather high mechanical strength. Polymer fibers meet these conditions.

Adsorption on the oil-absorbing materials from polymeric fibers proceeds by the principle of capillary condensation. This phenomenon is due to the influence of the surface curvature of the liquid meniscus on the vapor pressure of the substance over it. The relationship between vapor pressure above the meniscus and above the flat surface of the liquid is established by the Thompson's equation (equation 1):

$$p_0 / p = exp(2\sigma \cdot V_0 \cdot cos\theta / R \cdot T \cdot r), \qquad (1)$$

where σ is a surface tension of liquid; V_0 is a molar volume of liquid; r is radius of capillary; $cos\theta$ is a regional angle of wetting of surface of material by liquid.

Analysis of the Thompson's equation shows that capillary condensation is determined by the wetting of the capillary walls by liquid and its surface tension. The radius of the capillary depends on the porosity and the diameter of the fibers of the filter material.

The wetting of surface of polymer material by liquid was determined by direct measurement of drop under a microscope. The surface tension was determined by the method of separation of the ring from the liquid. Thin film of polymer from fusion is pre-applied on a ring. Measurement of regional angles of wetting showed that the organic compounds (hydrocarbons of normal structure, aromatic (benzene, toluene, xylene), alcohols of normal structure) spread on a material surface**,** i.e. $cos\theta$ is close to 1.

Table 2 gives the experimental data about wettability of polymeric materials by water. Table 3 gives the data about surface tension of hydrocarbons on a surface of polymers.

Table 2. Angle of wetting and surface tension on border "water – polymer".

Material	$cos\theta$	$\theta,^0$	σ, mN/m
Polyethylene (PE)	0.670103	47.925	68.715
Polypropylene (PP)	0.324324	71.075	66.533
Polyethylene terephthalate (PE TP)	0.724138	43.602	56.717

From the table 3 it is seen that the polypropylene is a hydrophobic material. Accordingly, polypropylene fiber preferably used to collect oil from a water surface, as polypropylene density is 0.9 – 0.91 g/cm^3 and it is lower than the density of water. PE is more hydrophobic than PP, but fibers from recycled PE TP have less mechanical strength than the fibers of recycled PP.

Table 3. Surface tension of hydrocarbons on a surface of polymers

Hydrocarbon	Polyethylene (PE)	Polypropylene (PP)	Polyethylene terephthalate (PE TP)
C_xH_{2x+2}, $X' = 3 \div 7$	0.775x +24.455	0.7606x+22.79	0.6602x+26.407
Benzene, toluene	1.6361x+29.267	2.7268x+22.723	1.6361x+28.359
Alcohols: C_2H_5OH, C_3H_7OH	1.09007x+29.449	1.2271x+30.904	1.2271x+31.994

where X' – number of carbon atoms in hydrocarbon molecule.

The data of table 3 also confirm that the polypropylene is a good material for the production of fibrous oil-absorbing materials not only for collecting oil and oil products from water surface, but also for use in the processes of filtration purification of water from oil products.

3. Results and considerations

Fiber with an average diameter of 70 μm was used in experiments for determination of oil-absorbing characteristics of polypropylene fiber; it was obtained by technology [4] from waste polypropylene grades of 21030 – 16 ÷ 21060 – 16. In the experiments the fiber density of samples (in free blow) was 150 ÷ 160 kg/m^3 and the porosity was 84 ÷ 86%.

The absorption capacity of the fiber samples for oil and oil products was determined by the gravimetric method according to the method of SShZhI–1202–96 of Institute of Petroleum of Chemistry Russian Academy of Sciences, Siberian Branch, Tomsk, Russia. Preweighed sample of the fiber was placed on a water surface with a layer of oil and oil products thickness with 3 ÷ 6 mm.

Then the sample with oil products was weighed, centrifuged at a factor of division 100 ± 3, again weighed and placed in capacity with oil products and water up to 50 times.

From table 4 it can be seen that the fibers with a small diameter have a greater sorption capacity. It is caused by surface and capillary forces of holding of hydrocarbons, which increase with decreasing of fiber diameter of the sorbent and the size of the pores between the fibers.

Table 4. Dependence of mass of absorbed oil from diameter of polypropylene fiber.

Type of hydrocarbons	Diameter of fibers, μm	Specific surface, m^2/g	Sorption capacity, g/g	Porosity, %
Collecting West Siberian Oil	70	0.057	4.03	87
Collecting West Siberian Oil	56	0.071	4.83	83.5
Collecting West Siberian Oil	42	0.094	9.3	80
Diesel fuel	70	0.057	4.01	87
Diesel fuel	56	0.071	5.5	83.5
Diesel fuel	42	0.094	9.6	80
The aviation kerosene	70	0.057	2.3	87
The aviation kerosene	56	0.071	5.6	83.5
The aviation kerosene	42	0.094	5.3	80

Table 5 presents the comparative data of filtration purification of water from oil by the fibers obtained by technology [4] from waste of polypropylene and fibers which were let out industrially such as sipron, which are widely used for filtration purification of water. The average diameter of the fibers of both types used in the experiment was 40 μm. Collecting West Siberian Oil was poured in water and an emulsion was prepared by multiple shaking of the container before disappearing of visible oil stains on the surface. The concentration of oil in water was 40 mg/l is the standard concentration of oil in water, which taken as a basis for designing of treatment facilities, for example, for carwashes. The filtration was finished at a fixed initial rate of reduction of gravity filtration on 10%, and in that moment the concentration of oil in the filtrate begun to rise sharply.

From the data presented in table 5 it is seen that the obtained fibers of polypropylene are more effective than sipron for purify the water from emulsified oil in mode of the gravity filtration.

Table 5. Results of filtration of mixture "water-oil" on polypropylene fiber and sipron.

Indicator	Fiber PP	Sipron
Density of fibers in column [kg /m^3]	158	168
Volume of emulsion passed through the column ([m^3/ kg] of fiber)	837	236
Gravity filtration rate at the start of process [m/hour]	97.0	95.5
Sorption capacity for oil ([kg] on 1 [kg] of sorbent)	3.57	0.53
Residual concentration of oil in the filtrate [mg/l]	0.45 ÷ 0.55	0.7 ÷ 1.8

4. Summary

It is established that polypropylene fiber with the greatest efficiency can be used as oil - absorbing materials during liquidation of emergency floods of oil and oil products on the surface of the water. It is preferable to use a fiber with a diameter of 40 – 70 μm for these purposes.

The fiber has high buoyancy and water repellency, well moistened by hydrocarbons, and production of them from waste provides a lower cost compared to fiber derived by traditional methods.

It is shown that fibers from polypropylene can be used in filtration processes for purification of oily wastewater. Polypropylene fiber has a greater sorption capacity and a higher degree of water purification in comparison with sipron fiber which is traditionally used for these purposes.

An additional advantage of polypropylene fiber is reusability in the mode "absorption-regeneration".

References

[1] Petrova A Y *et al* 2015 *Technical Physics* **60 (4)** 592–594
[2] Surzhikov A P *et al* 2009 *Nuclear Instruments and Methods in Physics Research, Section B* **267 (7)** 1072–1076
[3] Surzhikov A P *et al* 1999 *Russian Physics Journal* **42 (5)** 490–495
[4] Gorojankina G I *and* Pinchukova L I 2000 *Pipeline transport of oil* **4** 31–36
[5] Bordunov V V *et al* 2002 RF Patent 2179600 *Byull. Izobret.* **19**
[6] Galtseva O V *and* Bordunov S V 2014 *Testing. Diagnostics* **13** 68 – 72
[7] Galtseva O V *and* Bordunov S V 2015 *IOP Conf. Series: Materials Science and Engineering* **81 (1)** 112010 DOI: 10.1088/1757-899X/81/1/012062
[8] Bordunov S V *and* Galtseva O V 2016 *Journal of Physics: Conference Series* **671 (1)** 012009 DOI: 10.1088/1742-6596/671/1/012009

IOP Conf. Series: Materials Science and Engineering **132** (2016) 012004 doi:10.1088/1757-899X/132/1/012004

Thermo-Electromotive Force and Electrical Resistivity of Hydrogenated VT1-0 Titanium Alloy

A Lider[1], V Larionov[2], M Kroening[3] and V Kudiiarov[4]

[1]Head of Department, National Research Tomsk Polytechnic University, Tomsk, Russia
[2]Professor, National Research Tomsk Polytechnic University, Tomsk, Russia
[3]Professor, Saarland University, Saarbrucken, Germany
[4]Assistant, National Research Tomsk Polytechnic University, Tomsk, Russia

E-mail: lider@tpu.ru

Abstract. The method for measuring the structure transition of hydrogenated titanium from one state to another is suggested. The method is based on the comparison of thermo-electromotive force (thermo-emf), DC electrical resistance and the results of X-ray diffraction analysis. X-ray diffraction analysis is applied for identifying the quantity of defects in titanium structure. The authors have also identified the identical dependence of thermo-electromotive force and electrical resistivity on hydrogen concentration in titanium. The effect can be used for hydrogenated titanium structure control.

1. Introduction

Complex structural changes occur in titanium products during their operation. They suffer the formation of gas-saturated layers containing interstitial phases of oxygen, nitrogen and hydrogen with different concentration along depth. This complicates mechanical treatment, reduces metal plasticity, and reduces the performance of the components [1-3]. Gases are controlled and outgassed using different methods; protective media and coatings or vacuum heating are used [3, 4]. Due to this, sections which differ from one another are formed on titanium surface. These local structural defects lead to the destruction of components during their operation. If a submicrocrystalline structure is formed in titanium, this leads to the increase of capacity to accumulate hydrogen in titanium [5, 6]. It is caused by the increase of grain boundary length in submicrocrystalline samples in comparison to coarse-crystalline ones. The most well-known methods to control the presence of hydrogen and other gases in metals are the methods of microhardness measurement, X-ray diffraction and photometrical analysis [1, 2, 6, 7]. These are laboratory methods, and they are difficult to use for mass manufacturing inspection and are ineffective for detection of gas-saturated parts.

The present work focuses on the development of a technique for detecting defects in the microstructure of hydrogenated titanium by measuring thermo-emf and electrical resistivity using X-ray diffraction analysis.

2. Materials and Methods. Experimental Procedure

Annealed samples of commercially pure titanium (Fe-0.18; Si-0.1;C-0.07; N-0.04; O-0.12; H-0.01) with the dimensions of 10.0x5.0x1.0 mm and 100.0x28.0x1.0 mm were chosen for the experiment.

Hydrogen saturation was performed using Gas Reaction Controller according to Sieverts' method at the temperature of 300 – 600 °C, pressure of 0.6 and 2 atm. and time of hydrogenation of up to 300 minutes [4]. As a result of the treatment, the samples with the mass content of hydrogen of 0.0008 – 0.4 wt% were obtained. Hydrogen concentration in VT1-0 samples was determined using RHEN 602 (LECO) gas analyzer. Dislocation density in titanium was determined by broadening of X-ray lines using Shimadzu XRD-6000 diffractometer with Cu-K_α source. The analysis of phase composition was performed using PCPDFWIN and PDF-4+ databases and POWDER CELL 2.5 depth profile analysis software.

The measurement of titanium sample DC resistivity was performed by 4-probe method [7] using KEIHLEY INSTRUMENTS instrument control software. EMF was measured using T-3SP setup with increased pressing of electrodes to the sample.

3. Results

The following equation 1 was used to estimate the change in dislocation electrical resistivity taken relative to the unit of their density N_d in the unit of volume depending on hydrogen concentration in titanium [8]:

$$D = \frac{\rho_d}{N_d} = \frac{\hbar k_F \Omega_a Q}{n_s e^2},. \tag{1}$$

where k_F is the value of propagation vector on Fermi level, Ω_a is atomic volume, Q is electron transport cross section, n_s is the number of charge carriers per atom, e is the charge of electron, n_s is the number of charge carriers.

Density of dislocations N_d was determined using the Equation 2 from [9] based on X-ray diffraction analysis of titanium samples:

$$N_d = \pi\beta^2 ctg^2\Theta / 16b^2,. \tag{2}$$

where β is the broadening of X-ray lines caused by lattice microdeformation, θ is the angle that corresponds to the X-ray line peak, b is the Burgers vector. The resistivity of the samples was measured by probe method for each hydrogen weight concentration in titanium [10]. The increment of the resistivity connected with the unit of density of chaotically distributed dislocations $\Delta\rho_d / N_d$ was calculated as

$$\Delta\rho_d / N_d = \frac{3\pi^2 \hbar Q}{e^2 k_F^2}. \tag{3}$$

where N_d is dislocation density that was determined from X-ray measurements [9]. The increment of emf absolute coefficient equals to [11-13]

$$\Delta S = S_0 \frac{\Delta\rho_d}{\rho}\left(\frac{\Delta x}{x} - 1\right),. \tag{4}$$

where

$$\rho = \rho_0 + \Delta\rho_d; \rightarrow x = -\frac{k_F}{2}\left(\frac{d\ln}{dk}\rho_0\right)_{k=k_F}; \rightarrow \Delta x = -\frac{k_F}{2}\left(\frac{d\ln\Delta\rho_d}{dk}\right)_{k=k_F} = -\frac{k_F}{2Q}\left(\frac{dQ}{dk}\right)_{k=k_F} - 1.$$

Thus, the thermo-emf fraction change S is calculated as

$$\frac{\Delta S}{S_0} = \frac{3\pi^2 \hbar Q}{e^2 k_F^2 \rho}\left(\frac{\Delta x}{x} - 1\right). \tag{5}$$

and is totally dependent on the value of cross-section of electron scattering after interacting with defects, i.e. on the quantity and cross section shape of the latter. Such phenomenon is typical for the change of titanium alloy structure. According to Equation 5, the ratio $\Delta S/S$ is falling with the steepness depending on the value of Δx and scattering cross-section Q with distinctive inflection. In Figure 1, there is the inflection of emf dependence on hydrogen concentration in titanium, which is caused by titanium transition from one structural state to another. Further increase in hydrogen concentration does not change emf. Jump of emf can be explained by the formation of a chemical compound of hydrogen and titanium. Phase transition $\alpha \rightarrow \gamma$ is observed within the given hydrogen concentrations. With tangents drawn (as shown by numbers 1 and 2) to the given areas of diagram and a perpendicular dropped to the axis of distribution, the obtained value of hydrogen concentration in metal conforms to the concentration of titanium transition to another structural state.

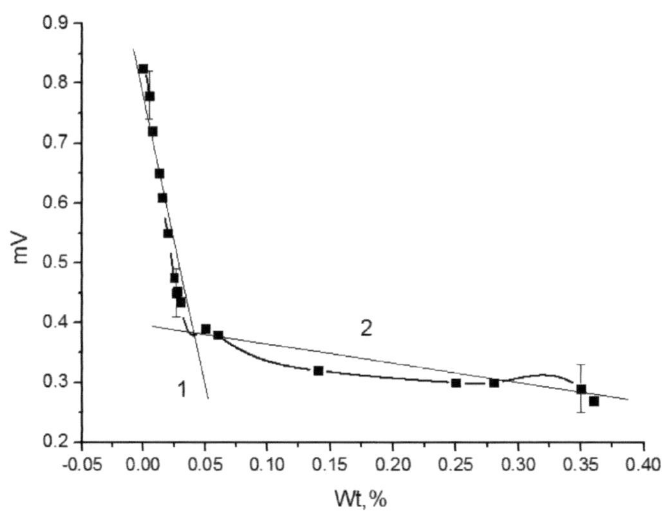

Figure 1. Dependence of thermo-emf on hydrogen concentration in titanium VT1-0.

The dependence of the number of defects in titanium on hydrogen concentration (Eq. 2) is presented in Figure 2.

There are two branches. With tangents drawn to each of the branches and a perpendicular dropped from the intersection of curves we obtain the hydrogen concentration in titanium corresponding to the change of electrical resistivity of the sample. Correlation of diagrams presented in Figures 1-2 gives the identical character of dependencies with a characteristic inflection point at hydrogen concentration of 0.05 wt%. The effect observed during the experiment can be used for evaluating titanium alloy transition from one state to another for controlling the state of the material. Accounting the pressing of electrodes to the tested sample allows implementing this method for laboratory and industrial control.

4. Conclusion

The suggested method for controlling the hydrogenated titanium structure is based on the simultaneous measurement of thermo-emf and electric resistivity. We have established two typical regions corresponding to the influence of hydrogen presence in titanium on the value of thermo-emf. The discovered indicative inflection of the dependence of thermo-emf on hydrogen concentration enables the determination of hydrogen concentration in titanium that coincides with the initiation of titanium alloy restructuring. We have shown that the transition from one structure to another is obvious even at hydrogen concentration in titanium of about 0.5 wt%. The correlation of structure and phase composition of titanium alloy as Ti H_2 : 4.04 wt% Ti H : 2.02 wt% Ti $H_{0.5}$: 1.01 wt% allows controlling the types of titanium-hydrogen compounds by measuring thermo-emf and using the plot thermo-emf vs. hydrogen concentration in titanium.

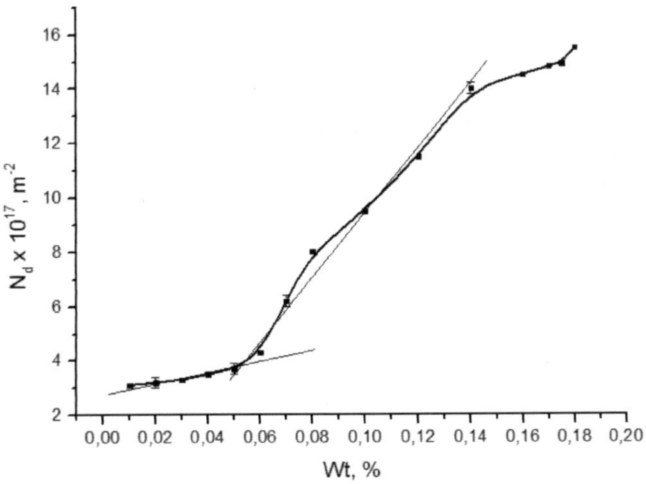

Figure 2. Dependence of number of defects on weight hydrogen content w_t in titanium alloy.

Acknowledgements
This study was partially financed by the Government of RF (Nauka grant no. 1524, theme 0.1325. 2014 and Federal Target Programme «Research and Development in Priority Fields of Russia`s Technological Complex Development for 2014 - 2020». Project unique identifier is RFMEFI57514X0048.

References

[1] Eliezer D *et al* 2000 Positive effects of hydrogen in metals *Materials Science and Engineering*. **A280** 220–224
[2] Eliaz N *and* Eliezer D 1999 An Overview of Hydrogen Interaction with Amorphous Alloys *Adv. Perf. Materials* **6 (1)** 5–31
[3] Kolachev B A *and* Talalaev D V 1993 Hydrogen technology of titanium alloys *Titan* **1** 43
[4] Kudiiarov V N *et al* 2014 Hydrogen Accumulation in Technically Pure Titanium Alloy at Saturation from Gas Atmosphere *Adv. Mater. Res* **880** 68–74
[5] Laptev R S *et al* 2014 Investigation of Defects Accumulation in the Process of Hydrogen Sorption and Desorption *Adv. Mater. Res.* **880** 134–138
[6] Evard E A *et al* 2005 Study of kinetics of hydrogen sorption and desorption from titanium *J. Alloys Comp.* **404** 335–338
[7] Kobeleva S P 2007 *Inorganic materials* **73 (1)** 60–67
[8] Karolik A S 2011 *Inorganic Materials: Applied Research* **4** 15–21
[9] Malychin D G *and* Korneeva V V 2010 The calculation of the X-ray diffraction of the field distortion of edge dislocations *Bulletin of the Kharkov University. Series Physical* **887** 115–119
[10] Larionov V V *et al* 2015 Eddy Current Analysis for Nuclear Power Materials *Adv. Mater. Res.* **1085** 335–339
[11] Voort V *and* Guot E P 1970 On the Electrical resistivity by scattering on Metallic Grain Boundaries *Phys. Stat. Solid* **38** 409–414
[12] Karolik A S *and* Luhvich A A 1994 Calculation of electrical resistivity produced by dislocation and grain boundnaries in metals *J. Phys: Condens. Mater.* **6** 873–886
[13] Karolik A S 1988 *The Physics of Metals and Metallography* **65** 463–469

IOP Conf. Series: Materials Science and Engineering **132** (2016) 012005 doi:10.1088/1757-899X/132/1/012005

Angular Velocity's Neural Network Observer of the Electric Drive of TVR - IM Type Implemented in Software Environment LabVIEW

L Kozlova[1], E Bolovin[1] and L Payuk[2]
[1]Assistant, National Research Tomsk Polytechnic University, Tomsk, Russia
[2]Senior Lecturer, National Research Tomsk Polytechnic University, Tomsk, Russia

E-mail: kozlovale@tpu.ru

Abstract. One of the common ways to manage a smooth starting and stopping of asynchronous motors are soft-start system. For this provision is necessary to use a closed speed asynchronous electric drive of tiristor voltage regulator - induction motor (TVR-IM) type. Using real sensors significantly increases the cost of installation and also introduces a number of inconveniences in the operation of the actuator. Observer has clear advantages that are created on artificial neural network. Creating a neural network observer in program graphic programming LabVIEW will allow to evaluate the speed of rotation of the asynchronous electric.

1. Introduction

Using of real sensors in the system of control of asynchronous electric drive of TVR-IM type technically and not economically justified. There are several advantages in the neural network based on which the observer was established: robustness, ease of establishment, there is no need to know the internal drive parameters, and others.

The article proposes to implement the neural network observer of angular velocity in the software environment graphical programming LabVIEW [1]. Currently artificial neural network (ANN) [2] is used in industries such as autonomous control, robotics, computers, and others.

This software product is useful in programming the sensors, graphical representations of various processes et al., created through a broad range of virtual instruments (VI) and functions for collecting, analyzing, displaying, and storing data. LabVIEW can visually identify and correct errors in the code. LabVIEW consists of three main parts: the front panel, block diagrams and subVI. The front panel is the user interface of the virtual instrument. To manage objects that are located on the front panel, using graphical representations of functions placed on the block diagram.

2. Experimental

Neural network observer consists of two main parts: the block of data preprocessing and neural network. The block diagram of such an observer, implemented in the software environment LabVIEW, is shown in figure 1, contains four subVI that act as a block of data preprocessing and a model of the observer. The block diagram contains four oscilloscopes. Block «Wait Until Next ms Multiple Function», with a constant of 0.00005, organize frequency of data collection.

IOP Conf. Series: Materials Science and Engineering **132** (2016) 012005 doi:10.1088/1757-899X/132/1/012005

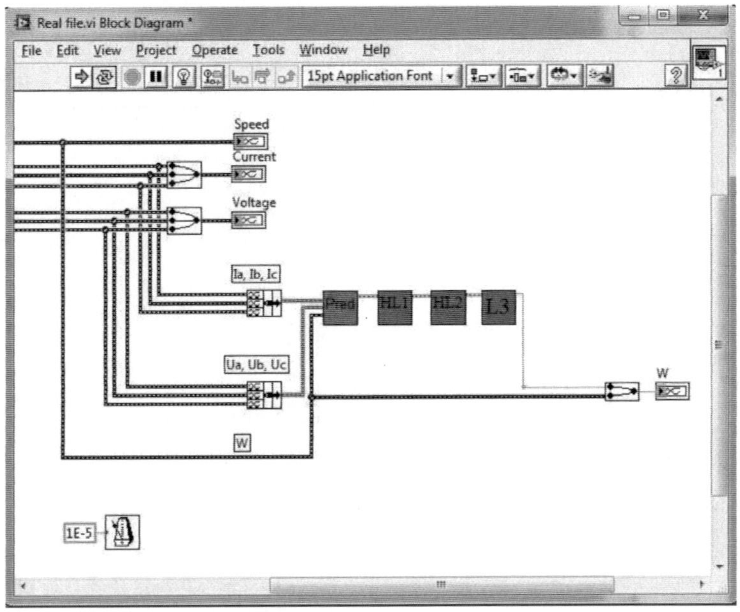

Figure 1. The block diagram of the simulation model.

On the block preprocessing data receives signals from the current and voltage sensors, which converts the signal and the delays in the polar coordinate system. The delay block uses only three current and three voltage:

$$I_M(t) = \sqrt{I_A(t)^2 + I_B(t)^2 + I_C(t)^2};$$

$$I_M(t - \Delta t) = \sqrt{I_A(t - \Delta t)^2 + I_B(t - \Delta t)^2 + I_C(t - \Delta t)^2};$$

$$I_M(t - \Delta 2t) = \sqrt{I_A(t - \Delta 2t)^2 + I_B(t - \Delta 2t)^2 + I_C(t - \Delta 2t)^2};$$

$$I_M(t - \Delta 3t) = \sqrt{I_A(t - \Delta 3t)^2 + I_B(t - \Delta 3t)^2 + I_C(t - \Delta 3t)^2};$$

$$U_M(t) = \sqrt{U_A(t)^2 + U_B(t)^2 + U_C(t)^2};$$

$$U_M(t - \Delta t) = \sqrt{U_A(t - \Delta t)^2 + U_B(t - \Delta t)^2 + U_C(t - \Delta t)^2};$$

$$U_M(t - \Delta 2t) = \sqrt{U_A(t - \Delta 2t)^2 + U_B(t - \Delta 2t)^2 + U_C(t - \Delta 2t)^2};$$

$$U_M(t - \Delta 3t) = \sqrt{U_A(t - \Delta 3t)^2 + U_B(t - \Delta 3t)^2 + U_C(t - \Delta 3t)^2},$$

where $I_A(t)$, $I_B(t)$, $I_C(t)$, $U_A(t)$, $U_B(t)$, $U_C(t)$ - a designation vectors of current and voltage.

Block diagram of the data preprocessing shown in figure 2, consists of such items as [1]:

1. Element «Unbundle Function» designed to separate the data from the block «Claster»;
2. Divided data from current and voltage sensors are supplied to unit «Square Function». This unit produces a function in squared;
3. «Compound Arithmetic Function» - a device designed to sum functions. On the output of this block are summed squares of the currents and voltages;
4. «Square Root Function» - device for finding a square root function;

IOP Conf. Series: Materials Science and Engineering **132** (2016) 012005 doi:10.1088/1757-899X/132/1/012005

5. «Feedback Node» - a device designed for organization a delay. To indicate the number of delays is necessary to trigger the properties of function and in the «Delay» indicate the number of delays.

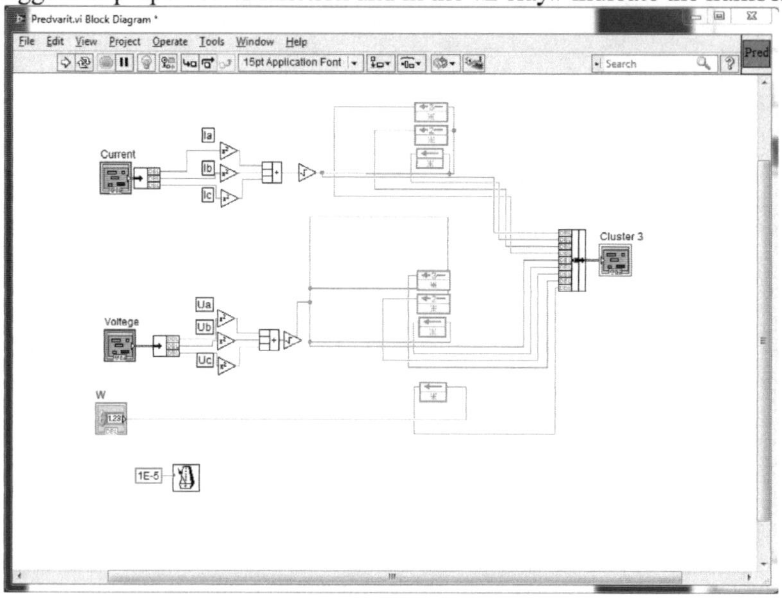

Figure 2. Block diagram of the data preprocessing.

After processing the data arrives at ANN, consisting of a set strictly ordered "elementary particles", which acts as a neuron. The network of neurons emulates the human brain. The mathematical model of a neuron can be described as follows [1, 3 – 14]:

$$S = \sum_{k=1}^{K} X_k \cdot W_k + b; \; Y = F(S),$$

where $X_1, X_2, ..., X_K$ – input signals of the neuron; $W_1, W_2, ..., W_K$ – synoptic weights of neurons; b – shift; $F(S)$ – activation function; Y – output signal neuron.

The developed neural network (figure 3) consists of a single input, two hidden layers and one output. The input layer consists of nine neurons, a first hidden layer – 7, the second – 27, output layer – from one neuron. Activation function of input layer is linear, hidden layers – tangential activation function and output layer – linear activation function.

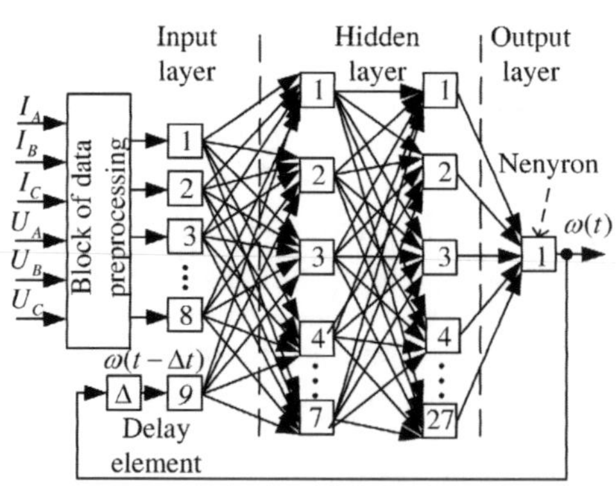

Figure 3. Block diagram of neural network observer of the angular velocity of the rotor.

IOP Conf. Series: Materials Science and Engineering **132** (2016) 012005 doi:10.1088/1757-899X/132/1/012005

The input layer of the neural network designed and implemented in LabVIEW software environment is shown in figure 4. Each subVI is the work of a single neuron with its synoptic weights and shifts. Block diagram of the neuron is shown in figure 5.

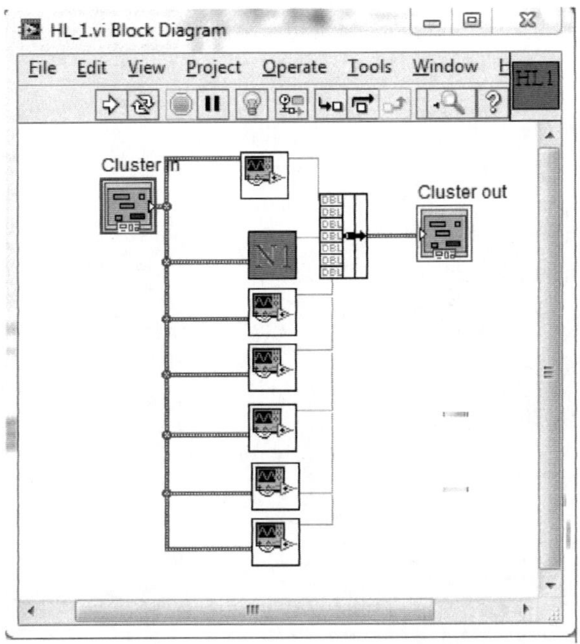

Figure 4. Block diagram of the first hidden layer.

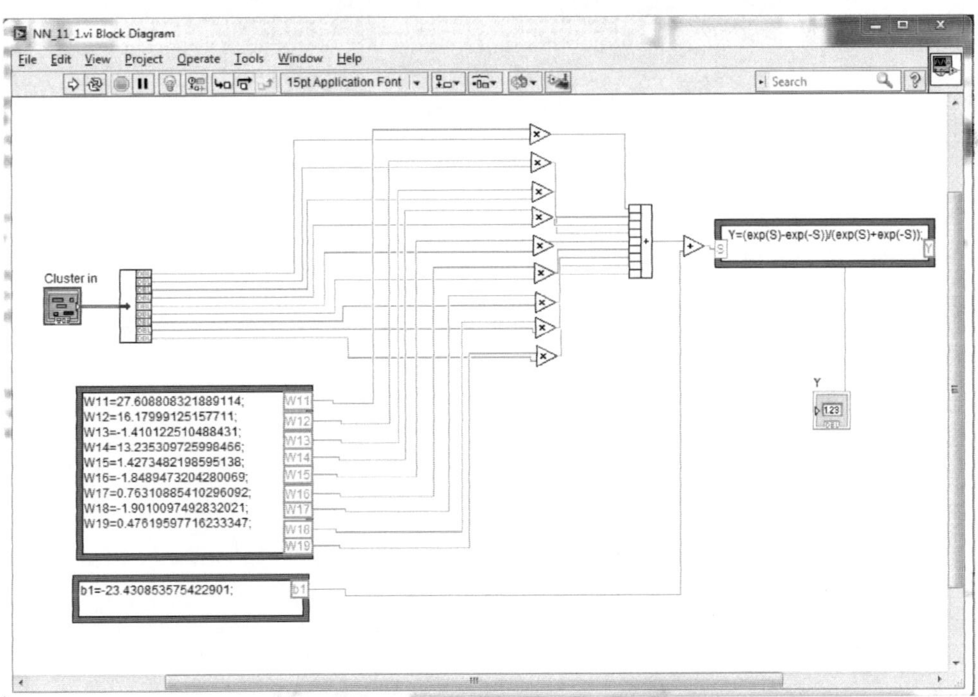

Figure 5. Block diagram of first neurons of the first hidden layer.

For the organization, neuron was used element «Formula Node» for writing formulas. With using this block the weights W, shifts b and tangential activation function can be arranged. In this way was organized each neuron of each layer of the neural network.

Have been obtained four waveforms to assess speed, voltage, current and comparative waveforms between the rotational speed of the motor shaft and its estimate implemented in the LabVIEW using oscilloscopes «Waveform Graph» (figure 6).

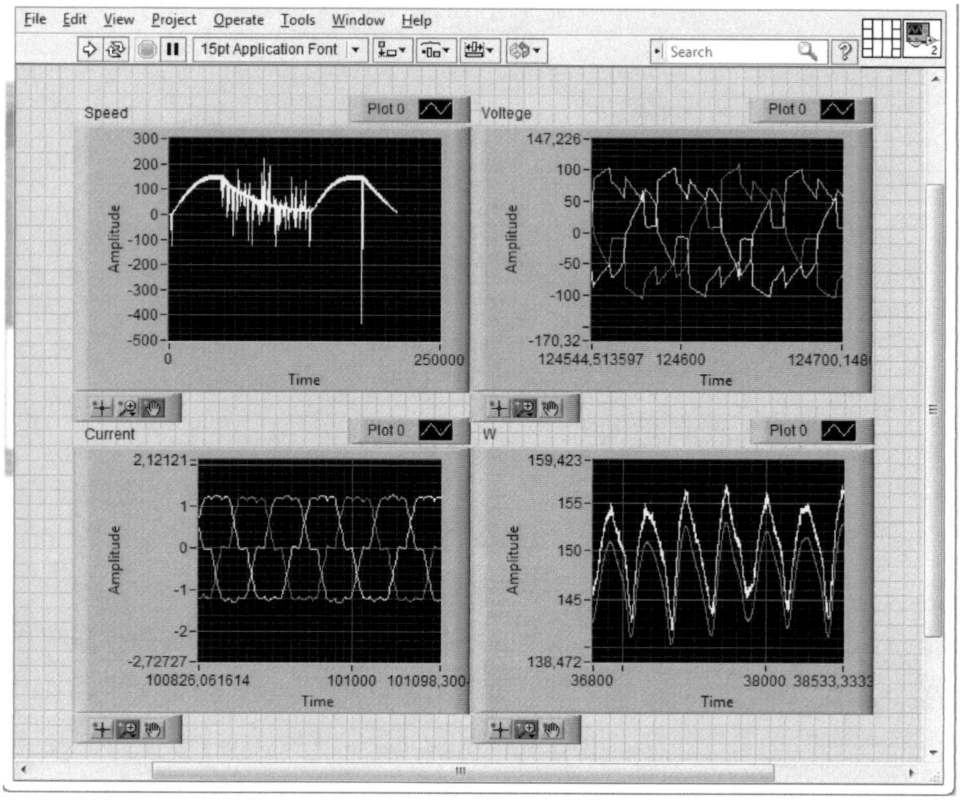

Figure 6. The front panel of the simulation model of neural network observer.

To assess the quality of the identification of the speed of rotation of the asynchronous electric drive of TVR-IM type integral criterion was used, calculated by the formula [14]

$$I_{\omega} = \frac{\int_{t_{start}}^{t_{stop}} |\omega(t) - \hat{\omega}(t)| dt}{\int_{t_{start}}^{t_{stop}} |\omega(t)| dt} \cdot 100\%$$

where t_{start}, t_{stop} – start and end times of integration, $\omega(t)$ – the angular velocity of rotation of the induction motor, $\hat{\omega}(t)$ - estimate the angular velocity of rotation of the asynchronous motor with different control modes are shown in table 1.

Table 1. The Integral criterion.

I_{ω}, %	Modes of operation of the electric drive		
	Start	Increased load	Load-shedding
	3,5	4,9	2,9

IOP Conf. Series: Materials Science and Engineering **132** (2016) 012005 doi:10.1088/1757-899X/132/1/012005

3. Summary

Based on table 1, created neyroemulyator of angular speed asynchronous electric drive of TVR-IM type produces adequate assessment despite the presence of noise, non-sinusoidal currents and voltages. The integral error is within 5%.

References

[1] National Instruments Corporation 2004 *LabVIEW PDA Module User Manual Information* **370624B-01** 30

[2] Bose B K 2001 *Modern power electronics and AC drive* (Prentice-Hall)

[3] Braslavskiy I Ya *et al* 2007 *Electrical Engineering* **11** 43–47

[4] Klepikov V B *et al* 1999 *Electrical engineering* **5** 2-6

[5] Payuk L A *et al* 2016 *Journal of Physics: Conference Series* **671** **(1)** 012044 DOI:10.1088/1742-6596/671/1/012044

[6] Kruglov V V *and* Borisov V V 2001 *Artificial neural networks. Theory and practice* (Hot line, Telecom)

[7] Shepherd G M *and* Koch C 1990 *The Synaptic Organization of the Brain* (Oxford University Press)

[8] Churchland P S 1986 *Neurophilosophy: Toward a Unified Science of the Mind-Brain* (Cambridge MA: MIT Press)

[9] Odnokopylov I G *et al* 2015 *Proc. Int. Siberian Conf. on Control and Communications* DOI: 10.1109/SIBCON.2015.7147249

[10] Bashkirov O A *et al* 1964 *Automation and Remote Control* **25** 629–631

[11] Lukichev D V *and* Usol'tsev A A 2005 *Scientific and Technical Gazette* **4** 97–102

[12] Kozlova L E 2015 *Scientific Bulletin NSTU* **1 (58)** 41–50

[13] Pritulov A M *et al* 2007 *Russian Physics Journal* **50** **(2)** 187–192 http://dx.doi.org/10.1007/s11182-007-0026-3

[14] Kozlova L E *and* Bolovin E V 2014 *Modern problems of science and education* **3**

The Level/Preventive Approach to Solving Technical and Economic Tasks as One of Directions in Development of the Enterprises of Oil Refining and Petrochemical Industry

R Damineva[1], **R Daminev**[2] **and O Karimov**[3]

[1]Head of Department, Ufa State Petroleum Technological University, Ufa, Russia
[2]Director of branch, Ufa State Petroleum Technological University, Ufa, Russia
[3]Associate Professor, Ufa State Petroleum Technological University, Ufa, Russia

E-mail: daminevarm@mail.ru

Abstract. In this work, as a tool of strategy generation in development of industrial enterprise, a methodological approach is considered, that allows the choosing the best path of development among many solutions of technical and economic tasks, taking into account both the interests of enterprises and of society as a whole.

1. Introduction

The industrial enterprises operating in the market environment are characterized by directing the available resources to the production sectors which would bring the maximum return. The perfect choice in effective use of production resources is the one, which possesses the lowest value of lost opportunities [1].

In developing the strategy, the senior managers of the company constantly face with the problems of further development, which are described by means of five well-known concepts of marketing: production rationalization, product development, intensification of selling efforts, marketing concept and societal marketing concept. In recent years, for the purpose of ensuring the competitiveness of production, a special importance was gained by the future-proof marketing concept.

The choice of a specific conception of the further development of industrial enterprise is predetermined by the emerging bottlenecks of the company. In particular, the reduction or the lack of demand for certain types of products and the weakening the competitiveness of the product or company. In general, this choice is predetermined by the negative impact of factors of macro- and micro-environment on the activity of the enterprise.

Let us consider in detail the reasons leading to the emergence of technical and economic problems in the industrial enterprise as well as the ways of their solution on the example of oil refining and petrochemical industry [2].

In the manufacture of petrochemical products, there are the following main production-branch stages: 1) crude oil production; 2) oil refining with the purpose of obtaining petroleum as fuel; 3) manufacturing of petrochemicals.

A common characteristic for the considered production stages is as follows:

1) general technological approach to equipment: heat exchangers, distillation columns, reactors, furnaces, etc.; 2) urgent need in use of auxiliary materials as catalysts, solvents and filters for

transformation of raw materials into semi-finished or finished products; 3) similar structure and composition of secondary production: maintenance, mechanical, transport and building services.

At each stage are built the manufacturing wastes are which in case of application of high-tech equipment can be used as raw material (trapped oil), either as sources of energy (heat and combustible waste energy) or as by-products (associated petroleum gas).

Let us consider the features of each production stage.

At the stage of crude oil production,one of the major objective problems in crude oil production is the decline in the quality of the extracted natural resource (high water cut, high content of sulfur compounds and solids, complicated geological and climatic conditions of production, etc.).

To subjective causes of the slowdown in crude oil production are as follows: 1) the policy of closing up the field geological exploration implemented in the last decades does not allow the entering of greenfields in the development. The geological exploration should be conducted so that the opened oil deposits would provide the required amounts of oil production for its subsequent processing at domestic enterprises. It is commonly known that in the last decade there has been a decline in investment activity in exploration of oil fields (up to 5% of all investments); 2) the privatization of former state-owned oil companies and creation of private oil extraction companies that implement the economic policy based on the known approach in economic science: maximum profit with minimum expenses (the maximum benefit taking into account minimization of expenses). The implementation of this principle is embodied in the increase in the share of oil exports and the shortage in the share of oil supplied to the domestic market. The peculiarity of the Russian Federation is a significant underutilization of production capacity of oil refineries.

Thus, the economic situation in oil production of the Russian Federation is determined by the political situation.

To resolve this problem, it is necessary that the income from the use of mineral resources would have made a greater contribution to the formation of revenue part of state budget. For this purpose, it is necessary to develop a legislative framework by means of nationalization of mineral resources or by setting quotas on the export of oil.

Dynamics of mining fuel and energy resources in the Russian Federation is shown in Table 1 [6].

Let us consider the following production-branch stages – crude oil refining and petrochemistry - and select the following groups of problems faced by enterprises of oil refining and petrochemistry:

1. Decrease in gross profit; 2. Increase in production costs; 3. Decrease in competitiveness of the products on quality parameters compared with international manufacturers.

The causes of above problems are identical to the reasons for existence of cyclical development of the economy. As is known, there are external and internal reasons for economic fluctuations.

The external reasons include:

- wars, revolutions and other political events;
- the discovery of large deposits of gold, uranium, oil and other valuable resources;
- the development of new territories and the consequent migration of the population, fluctuations in the population of the globe;
- powerful breakthroughs in technology and inventions, which allow drastical changes in the structure of social production.

The internal causes of cyclical development of the economy are as follows:

- the actual service life of fixed capital;
- personal consumption, whose reduction or increase affects the volume of production and employment;
- investing, i.e. investment of capital in the expansion or modernization of production as well as in creation of new jobs;
- economic policy of the state, expressed in the direct and indirect effects on demand and consumption.

Table 2 shows the dynamics of the major types of oil products.

Thus, the entity can have either separate local technical and economic problems, or in most cases a set of interdependent shortcomings in production activity.

The resolution of problems identified in the framework of traditional technologies is not always beneficial for the company. Therefore it is necessary to consider also other ways of solving technical and economic problems.

Table 1. Extraction of the fuel and energy natural resources, million tons.

Type of natural resource	1990	1995	2000	2001	2002	2003	2004	2005	2006	2007	2008	2009	2010	2011	2012	2013	2014
Coal, million tons	395	263	258	270	256	277	282	299	310	314	329	301	322	335	357	351	356
Peat, million tons	73.0	13.5	4.1	4.6	3.0	1.5	2.0	1.9	1.7	1.7	1.2	1.2	1.3	1.5	1.2	1.5	1.1
Oil including gas condensate, milliontons	516	307	324	348	380	421	459	470	481	491	488	494	505	512	519	522	525
Gas combustible natural (natural), milliardm3	641	595	584	581	595	620	633	641	656	653	666	583	651	671	655	668	639

Table 2. Production of main types of oil products, million tons

Product types	1990	1995	2000	2001	2002	2003	2004	2005	2006	2007	2008	2009	2010	2011	2012	2013	2014
Primary oil refining	298	182	173	179	185	190	195	208	220	229	237	237	250	258	250	258	272
Motor gasoline	40.9	28.1	27.2	27.6	29.0	29.3	30.5	32.0	34.4	35.,1	35.,6	35.8	36.0	36.,7	36.0	36.7	38.2
Diesel fuel	76.2	47.3	49.2	50.2	52.7	53.9	55.4	60.0	64.2	66.3	68.9	67.2	70.0	70.3	70.0	70.3	69.4
Heating oil (total output)	94.0	61.4	48.2	50.3	54.2	54.6	53.6	56.7	59.3	62.7	63.9	64.6	69.6	73.2	69.6	73.2	74.4

2. Methods

We propose a level/preventive approach of solving technical and economic problems.

Implementation of level approach is as follows. As it is known, the technological structure of enterprises of the industries considered is interconnected: the products of some production sites, process plants and workshops act as components of raw materials, semi-finished products, auxiliary materials, fuel and energy resources for other production stages.

The same situation is with by-products and wastes, but at larger spatial scale. It can appear that the by-products or production waste of some enterprises act as components of material resources for other companies. Not always the financial and production capabilities of entity allow the resolving of technical and economic problems sufficiently profitable.

Therefore, at the first stage we propose to identify the companies which will be interested in buying of by-products or wastes of production formed at the particular plant. To save the transport costs, the search for possible partners should be started from the regional level (Fig. 1). In process of expansion of radius of search of partners, the probability of solving technical and economic task increases.

IOP Conf. Series: Materials Science and Engineering **132** (2016) 012006 doi:10.1088/1757-899X/132/1/012006

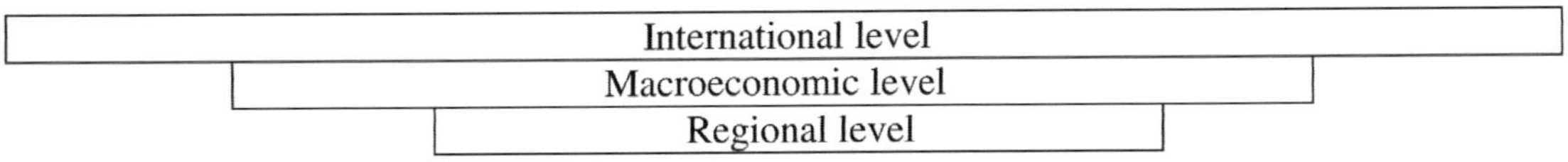

Figure 1. The level method for solving technical and economic problems.

For the successful implementation of this model, a data bank for enterprises is to be created. A tentative list of baseline information contributing the creation a data bank is shown in table 3.

Table 3.The baseline information to create a data bank.

Name of the company	Location of the company	Nomenclature and range of products	Formulation of technical-economic task	Technical parameters of the products underlying the technical-economic task

If in the result of analyzing the data bank it is not possible to obtain the desired solution of the problem in the plant, then it is necessary to proceed to the second stage – preventive approach in solving technical and economic problems of the plant.

The preventive approach consists in point wise application of the most modern high-tech technologies to solve specific problems of industrial enterprise. For example, the teachers of Sterlitamak branch of Ufa State Petroleum Technological University develop the technical solutions for the use of alternative methods of influence on the technological environment (microwave radiation) in the production of isoprene rubber at processing of oil sludge [3–5].

The above calculation of integral indicators of effectiveness of these design solutions is justified by their economic feasibility. The success will consist in applying the most effective energy-saving technologies, focusing on solving specific technical problems. For successful implementation of preventive approach it is important to understand: the more specific the problem will be put (i.e. the technical task is to be narrowed to the maximum possible level), the less is the cost.

The optimal solution of technical and economic tasks can consist solely in the unification of existing methods with the economic evaluation of the cost of level and preventive approaches. For economic evaluation of the solutions considered, the known method of economic evaluation of efficiency of investment projects can be used. In the presence of both level and preventive solutions of technical and economic tasks, as the best one will be considered the project, which will have the highest value of net discounted income and profitability index with sufficient underlying strength.

3. Conclusion

Thus, the use of level/preventive approach in solving the technical and economic problems will contribute to working out the strategy of enterprise development, whose successful implementation will allow the optimal use of the limited financial resources of the company. By doing so, the macro – and microeconomic interests will be favorably combined.

References

[1] Izotova E V 2008 *Int. Workshop on Computer Science and Information Technologies (Antalya)* 180–198
[2] Damineva R M 2013 *Improvement of economic and financial activity and organizational structure of industrial enterprise* (Moscow: Approbation Inc.)
[3] Karimov O K *et al* 2013 *World Applied Sciences Journal* **24(3)** 320–325
[4] Karimov O K *et al* 2013 *Middle East Journal of Scientific Research* **18(2)** 127–130
[5] Yunusov D S *et al* 2011 *International Polymer Science and Technology* **38(5)** 37–38
[6] Information on http://www.gks.ru

IOP Conf. Series: Materials Science and Engineering **132** (2016) 012007 doi:10.1088/1757-899X/132/1/012007

Integrative Modeling of Electrical Properties of Pacemaker Cardiac Cells

M Grigoriev[1] and L Babich[2]

[1]Engineer, National Research Tomsk Polytechnic University, Tomsk, Russia
[2]Student, National Research Tomsk Polytechnic University, Tomsk, Russia

E-mail: mishatpu@mail.ru

Abstract. This work represents modeling of electrical properties of pacemaker (sinus) cardiac cells. Special attention is paid to electrical potential arising from transmembrane current of Na^+, K^+ and Ca^{2+} ions. This potential is calculated using the NaCaX model. In this respect, molar concentration of ions in the intercellular space which is calculated on the basis of the GENTEX model is essential. Combined use of two different models allows referring this approach to integrative modeling.

1. Introduction

The cardiac muscle comprises typical and atypical pacemaker cardiac cells which form the cardiac conduction system. Excitation from the atrioventricular node under normal conditions is possible only in one direction. The most important condition for assurance of cardiac cells functioning is anatomical continuity of the whole conduction system. Topography of pacemakers and ionic mechanisms of ventricular connection and ventricular conduction system is not fully understood nowadays.

A great number of interrelated electrochemical processes with involvement of cardiac cells are conducted continuously in the heart, as a result of which action potential leading to cardiac contraction is generated. Disorder of this complex system leads to a wide range of cardiovascular diseases which are the main factor of mortality in advanced economies. Timely diagnostics of potentially adverse states might remedy the situation [1–4]. However, there are certain tasks which have not been settled yet by the modern technologies of experimental diagnostics, e.g. simultaneous measurement of dynamics of several transmembrane currents. There is a need to develop the area of mathematical modeling and comparative studies of electrophysiological arrangement of atrioventricular connection and ventricular conduction system cells.

The mathematical model will make it possible to explore general rules of cardiac muscle functioning more deeply and determine the disorder mechanism of the electrical and contractile function of cardiac cells.

2. Analysis of technologies

Conceptual models of the cardiac function have been developed since the 50s of the 20th century. Models of this class allow obtaining the qualitative idea of the studied phenomenon. They have relatively minor computational complexity, and their application is often not limited to the field wherefore they were elaborated. The example thereof is the FitzHugh-Nagumo model [5] which is widely used not only for description of cardiac electrical activity, but also for other self-oscillating systems.

So called "detailed models", e.g. the family of Hodgkin-Huxley models [6], are applied in order to obtain qualitative estimates. Models of this class ensure design accuracy on the level of accuracy of experimental procedures which conditions their reasonable practicality. On the other hand, their resource intensity is substantially higher which is conditioned by complexity of the modeled system.

Computational complexity may be reduced significantly as part of the integrative approach that lies in synthesis of some relatively simple models describing different aspects of the studied system. Integrative modeling is most extensively used in the Physiome project [7], the purpose whereof is to build the complete human body model.

The most common models of intracellular ion concentration of ions in cardiac cells use calcium, sodium and potassium. Calcium concentration is important, and the majority of models elaborated since 1977 include at least some basic dynamics of intracellular calcium. Intracellular concentrations of sodium and potassium are easier to represent and updated only proportionally to the sum of transmembrane currents associated with each ion. Intracellular calcium models reproduce sophisticated dynamics of calcium inside the cell on some level. The increased level of intracellular calcium initiates the cell contraction process. Relatively small calcium influx through *L*-type calcium channels in the membrane raises calcium concentration in cytosol by approximately ten times. Calcium together with other cell compounds initiates the contraction process.

Cardiac muscle excitability arises under the influence of various stimuli, such as chemical, electrical, thermal ones etc., which condition the ability to shift to the excitation state. This phenomenon is based on negative electrical potential which arises at the primarily excited area. Functional cardiac cells in the excited tissue state have the polarized membrane. It has the positive charge on the outside and the negative one on the inside. Such a state is associated with different concentration of Na^+ and K^+ on both sides of the membrane. Different membrane permeability of cells plays a certain role for these ions. Na^+ ions do not penetrate through the cardiac cell membrane at rest. Potassium ions penetrate it only partially. Diffusion of K^+ ions escaping the cell increases the surface positive charge. Furthermore, the inner side becomes negatively charged. Sodium ions enter a cell under the influence of any stimulus. At this time the negative charge occurs on the cardiac cell membrane surface, and potential reversal is ensured. The amplitude of action potential of cardiac cells comprises around 10 mV and more. In its turn, this potential depolarizes membranes of adjacent cells. Eventually, they obtain their own action potentials, i.e. propagation of excitation occurs in other cardiac muscle cells as well.

Action potential of a functional cardiac cell is many times longer than that in skeletal muscles. Cardiac muscle cells do not get excited by the further stimuli during the period of development of action potential. This feature is essential for function of the heart as an organ. It is related to the fact that the cardiac muscle is able to draw only one action potential and can respond to repeated excitations only once. These phenomena create conditions for heart beating [6–8].

3. Modeling

Integrative modeling is also applicable to pacemaker cardiac cells [9, 10]. One of the most important characteristics of these cells is transmembrane voltage which is characterized by the ordinary differential equation as part of the model [11]:

$$\frac{dV}{dt} = -\frac{I_{Na} + I_{Ca,T} + I_{Ca,L} + I_K + I_f + I_B + I_{NaK} + I_{NaCa} + I_{CaP}}{C_m} \quad (1)$$

where $V(t)$ is the voltage generated by a cardiac cell; C_m is the electric capacity; symbol I with different subscripts represents the concentration of relevant ions:

I_{Na} is the sodium current, $I_{Ca,T}$, $I_{Ca,L}$ are potassium currents T of L types, I_{CaP} is the calcium pump, I_{NaK} is the Na, K pump, I_{NaCa} is the exchanger, I_f is the activated current, I_B is the background current, and I_K is the potassium current.

Nowadays, experimental data represented by single cellular channels show that opening and closing of channels are described by random processes. As a cell has many functioning channels, models given below characterize the average ion concentration.

It is apparent that transmembrane current depends both on the ion concentration in the intercellular space and its geometry. These dependencies are reflected in the model [12] which represents the partial differential system:

$$\frac{\partial C_p}{\partial t} = -\frac{F_p L}{V_p}\frac{\partial C_p}{\partial x} - \frac{PS_g}{V_p}(C_p - C_{isf}) - \frac{PS_{ecl}}{V_p}(C_p - C_{ec}) - \frac{G_p}{V_p}C_p + D_p\frac{\partial^2 C_p}{\partial x^2} \quad (2)$$

$$\frac{\partial C_{RBC}}{\partial t} = -\frac{F_{RBC} L}{V'_{RBC}}\frac{\partial C_{RBC}}{\partial x} - \frac{PS_{RBC}}{V'_{RBC}}(C_{RBC} - C_p) - \frac{G_{RBC}}{V'_{RBC}}C_{RBC} + D_{RBC}\frac{\partial^2 C_{RBC}}{\partial x^2} \quad (3)$$

$$\frac{\partial C_{ec}}{\partial t} = -\frac{PS_{ecl}}{V'_{ec}}(C_{ec} - C_p) - \frac{PS_{eca}}{V'_{ec}}(C_{ec} - C_{isf}) - \frac{G_{ec}}{V'_{ec}}C_{ec} + D_{ec}\frac{\partial^2 C_{ec}}{\partial x^2} \quad (4)$$

$$\frac{\partial C_{isf}}{\partial t} = -\frac{PS_g}{V'_{isf}}(C_{isf} - C_p) - \frac{PS_{eca}}{V'_{isf}}(C_{isf} - C_{ec}) - \frac{PS_{pc}}{V'_{isf}}(C_{isf} - C_{pc}) - \frac{G_{isf}}{V'_{isf}}C_{isf} + D_{isf}\frac{\partial^2 C_{isf}}{\partial x^2} \quad (5)$$

$$\frac{\partial C_{pc}}{\partial t} = -\frac{PS_{pc}}{V'_{pc}}(C_{pc} - C_{isf}) - \frac{G_{pc}}{V'_{pc}}C_{pc} + D_{pc}\frac{\partial^2 C_{pc}}{\partial x^2} \quad (6)$$

where F_p is the flow of dissolved substance contained in plasma; PS_{RBC} is the surface area of red blood cells; PS_g is the surface area for passive transport through clearances or gaps between adjacent endothelial cells; PS_{ecl} is the clearance area for endothelial cells; PS_{eca} is the contact surface area; PS_{pc} is the surface area of parenchymal cells (myocytes); G_{reg} is the local ion rate; D_{reg} is the diffusion coefficient in the axial direction of the V_p area; V_p is the intracapillary plasma volume; V'_{RBC}, V'_{ec}, V'_{isf} and V'_{pc} are virtual volumes of red blood cells, endothelial cells, intercellular space and cardiac cells respectively.

The model (2)-(6) with natural initial and boundary conditions allows evaluating the ion concentration on the outside of cardiac cells. Modeling results are presented on figure 1.

Figure 1 shows modeling time from 0 to 10 seconds along the horizontal axis and the standardized ion concentration in dimensionless units along the vertical axis. It is obvious from this figure that concentration rises sharply starting from the 4th second which corresponds with model preset parameters. Concentration fades out starting from the 5th second, thereby showing the ion rate in the transmembrane current. Voltage of this current calculated according to the model [11] is given on figure 2.

Figure 2 shows the modeling time (0-10 seconds) along the horizontal axis and the standardized value of electric voltage generated by a cardiac cell in dimensionless units along the vertical axis. A single beat in the beginning of the 2nd second is caused by initial values of model parameters. Voltage fluctuations corresponding with the beginning of the transmembrane current occur from the 4th second (figure 1). The amplitude of these fluctuations grows smoothly at first, and then decreases sharply starting from the 8th second. This mode of fluctuations corresponds fully with the local ion rate in the intercellular space (figure 1).

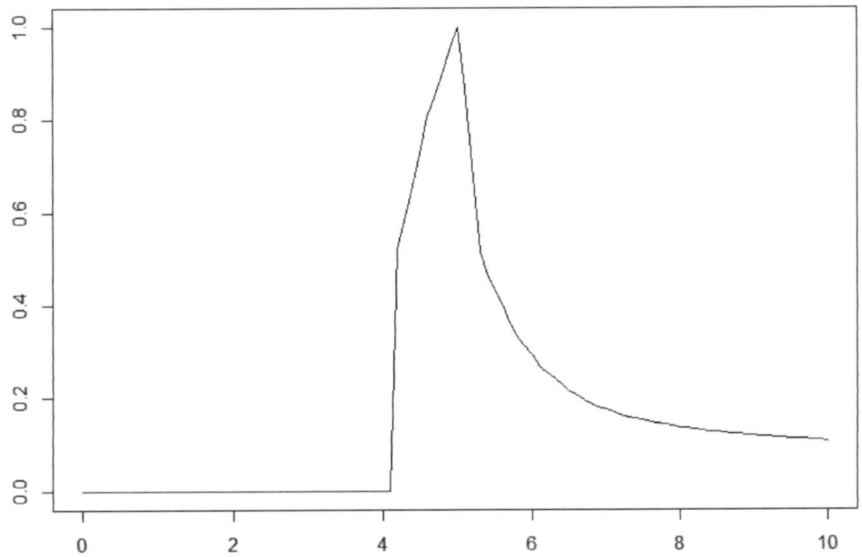

Figure 1. Dependence between ion concentration on the outside of a cardiac cell and time.

The specified results were obtained on the office computer with CPU frequency of 3.2 GHz and RAM of 8 Gb under control of the 64-bit operating system of the Windows family. Calculation time comprised approximately 20 seconds for each model. Thus, the integrative approach allows obtaining qualitative estimates of electrical activity of cardiac cells under random conditions. At the same time, fewer computation capacities are required as opposed to detailed models.

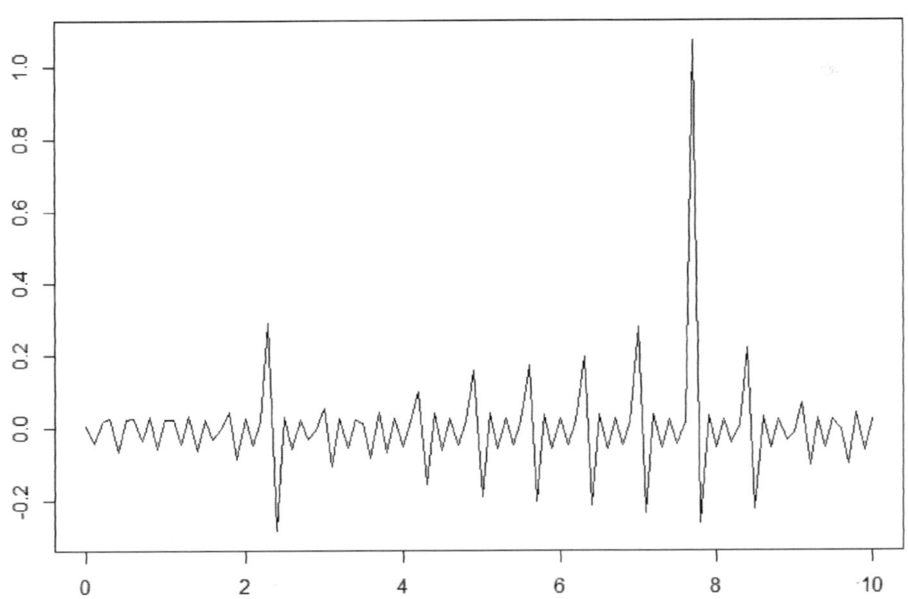

Figure 2. Dependence between cardiac cell electric voltage and time.

4. Conclusion

It is obvious that cardiac cells represent quite a sophisticated contractile "machine". Its arrangement principles and behavior may be predicted only using complex mathematical modeling. At present there are certain developments, and various mathematical models which describe subsystems of the cardiac muscle/cardiac cells with different degree of detail are elaborated. Models of passive mechanical properties of cardiac cells, models of their active mechanical behavior as well as the model of electrical activity of myocytes are described. However, there is still a considerable lack of integrative models which characterize the existing subsystems with account of direct links and feedback between them.

The mathematical model not only allows investigating rules of cardiac muscle behavior in the active mode thoroughly, but also enables to determine the disorder mechanism of the electrical and contractile function of cardiac cells which accompany different cardiac pathologies, as well as elaborate and implement new methods of treatment of various disorders. Possibilities of integrative methods for clinical practice are demonstrated by examples of analysis of arrhythmia mechanism and treatment methods which arise in case of calcium overload of cardiac cells that happens in many heart diseases. This model makes it possible to predict the role of mechanical and electrical links in pathogenesis of cardiac arrhythmia [13].

Acknowledgement

The research was financially supported by the Federal Target Program "Research and development on priority directions of scientific-technological complex of Russia in 2014-2020 years», project No. 14.578.21.0032 "Development of an experimental model of the hardware-software complex for noninvasive recording of the heart micropotentials in a broad band without filtering and averaging in real time in order to detect the early signs of sudden cardiac death". The unique identifier of the contract: RFMEF157814X0032.

References

[1] Grigoriev M *and* Turushev N 2015 *Applied Mechanics and Materials* **770** 526–530
[2] Avdeeva D *et al* 2014 *Biology and Medicine* **6** 1–6
[3] Grigoriev M *and* Turushev N 2014 *Advanced Materials Research* **1040** 928–932
[4] Payuk L A *et al* 2016 *Journal of Physics: Conference Series* **671** **(1)** 012044 DOI:10.1088/1742-6596/671/1/012044
[5] FitzHugh R 1961 *Biophysical J.* **1** 445–466
[6] Solovyova O *et al* 2002 *The J. of Physiology* **544** 22–23
[7] Hannes T *et al* 2008 *J. Electrocardiol.* **41** 562–566
[8] Avdeeva D K et al 2014 *Biology and Medicine* **6 (2)** BM–025 1–5
[9] Nikitina L V *et al* 2008 *Biochemistry (Moscow)* **73(2)** 178–184
[10] Bradd A D *et al* 2012 *Conf. Proc. Eng. Med. Biol. Soc. (San Diego, CA)* 195–198 DOI: 10.1109/EMBC.2012.6345904
[11] Demir S S *et al* 1994 *Am J Physiol* **266** 832–852
[12] Bassingthwaighte J B *et al* 2006 *Phil Trans Roy Soc: Mathematical* **364** 1423–1442
[13] Herrera-Valdez M A *and* Lega J 2011 *J. Theor. Biol.* **270** 164–176

Generation and Analysis of Wire Rope Digital Radiographic Images

S Chakhlov[1], P Anpilogov[2], A Batranin[3], S Osipov[4], Sh Zhumabekova[5] and I Yadrenkin[6]

[1]Head of Laboratory, National Research Tomsk Polytechnic University, Tomsk, Russia
[2]Expert, LLC "Tepromes", Tomsk, Russia
[3]Junior Researcher, National Research Tomsk Polytechnic University, Tomsk, Russia
[4]Leading Researcher, National Research Tomsk Polytechnic University, Tomsk, Russia
[5]Student, National Research Tomsk Polytechnic University, Tomsk, Russia
[6]Assistant, Tomsk State University of Architecture and Building, Tomsk, Russia

E-mail: chakhlov@tpu.ru

Abstract. The paper is dealt with different structures of the digital radiographic system intended for wire rope radiography. The scanning geometry of the wire rope is presented and the main stages of its digital radiographic image generation are identified herein. Correction algorithms are suggested for X-ray beam hardening. A complex internal structure of the wire rope is illustrated by its 25 mm diameter image obtained from X-ray computed tomography. The paper considers the approach to the analysis of digital radiographic image algorithms based on the closeness of certain parameters (invariants) of all unit cross-sections of the reference wire rope or its sections with the length equaling to the lay. The main invariants of wire rope radiographic images are identified and compared with its typical defects.

1. Introduction

For the last few decades, non-destructive testing techniques have been rapidly developed, including those applied for wire ropes. NDT techniques used for detecting manufacturing and operational defects of wire ropes include various optical, magnetic, ultrasonic, electrical, and other methods [1–3]. Wire ropes with large diameter are the most complicated test objects for NDT testing. Such objects can be tested by classical radiography which has a number of disadvantages, namely: the use of argentum-containing radiographic films that results in a high cost of testing and the extremely low performance due to the multi-stage and long-term formation of shadow images. Since the 80s of the past century, the different implementations of digital radiography have been rapidly developed [4–7] that are characterized by high performance, sensitivity, and the possibility of applying various algorithms of image processing. Radiometric detectors provide the generation of radiographic images having the geometrical resolution close to classical radiography. And a wider classification of high-performance X-ray sources stipulate for the use of radiography in NDT of wire ropes possessing large diameters [8–11].

2. Structures of the digital radiographic system

The digital radiographic image of a wire rope is generated by its scanning with a narrow X-ray beam produced by a slit collimator. Two scanning variants are possible. The first implies the wire rope movement relative to an X-ray generator-detector static system. In the second variant, this system moves relative to a wire rope in a static condition.

Wire rope scanning can be carried out by two ways: discrete and continuous. In discrete scanning, the mutual movement of the test object and the X-ray generator-detector static system is performed by fixed steps Δl. In continuous scanning, the line is formed during the time interval Δt. The parameters of discrete and continuous scanning relate to each other as $\Delta l = V\Delta t$, where V is the velocity of the wire rope movement relative to the X-ray generator. Radiographic image formation using pulsed X-ray sources can be performed solely by the integration method of radiation recording, while in a continuous X-ray source, both the integration and counting methods are possible. The counting method includes different implementations, namely: counting method proper; time; spectral.

3. Layout geometry of scanning a wire rope

Figure 1 presents the cross-sectional view of the digital radiographic system and continuous scanning with a narrow X-ray beam similar to that one described in the work of Osipov et al. [10]. The test object, namely the wire rope, can be represented by a steel cylinder with radius R. For convenience, let us assume that all line-scan detectors are oriented towards the X-ray source.

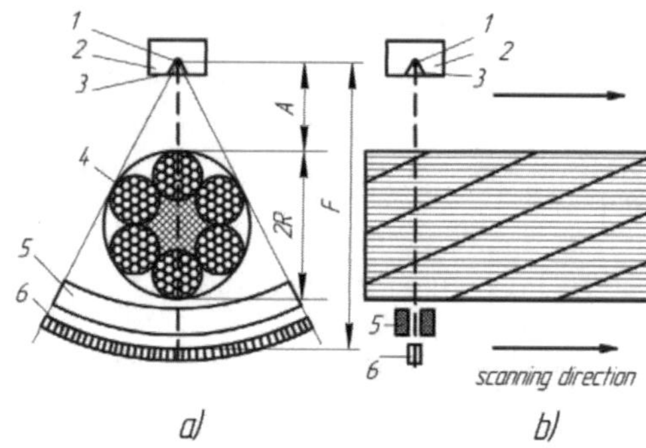

Figure 1. Cross-sectional view of digital radiographic system:
(*a*) normal to the scanning direction; (*b*) in the scanning direction.
1 – X-ray source; *2* – biological protection unit; *3* – embedded slit collimator;
4 – test object; *5* – slit collimator; *6* – detector array.

Single radiometric detectors in the array have the cross-sectional dimension of $a \times a$ and are located at distance F from the X-ray source. The additional properties of the digital radiographic system are the minimum distance A from the X-ray source to the test object and the arc length l which touches frontal surfaces of detectors. The original digital image I is $N \times M$ size matrix, where N is the number of pixels in length and $N = l/\Delta l = l/(V\Delta t)$; M is the number of detectors in the array. Let us assume that $M \ll N$. In case the detectors adhere to each other, the geometrical parameter l can be obtained from the formula

$$l \approx Ma \tag{1}$$

In radiographic systems based on pulsed X-ray sources, M is selected with a glance to the necessity for the use of end detectors in the capacity of reference channels intended for calibration

$$M > \operatorname{int}\left(2F \arcsin\left(R/(R+A)\right)/a + 0.5\right) + 2k + 2, \tag{2}$$

where R is the radius of the rope; k is the number of end detectors in the array that excludes the hypothetical shadowing of line-scan detectors by the test object.

4. Generation of digital radiographic images

Let us assume that detector m, m=1…M has two parameters, namely k_m and σ_m, i.e. the amplification coefficient and the inherent noise level, respectively. The wire rope is scanned by a narrow radiation beam produced by a slit collimator with the maximum energy E_{max}. The X-ray source is not isotropic. Therefore, it is necessary to reduce the influence of individual detector properties on the quality of final images. This can be achieved by both black and white calibration. Black calibration measurements are carried out at switched off X-ray source I_b, whilst white calibration measurements are carried out at switched on X-ray source I_w and without a test object. The original image I is formed with the test object. The number of lines N_b and N_w in images I_b and I_w is significantly lower than N.

By means of the analog-to-digital coder, a set of analog signals (images) I_b, I_w, I transforms to a set of digital signals (images) J_b, J_w, J. This process is described by equations

$$J_{nm} = [I_{nm}/\Delta I],\ n = 1 \div N,\ m = 1 \div M;\ J_{bnm} = [I_{bnm}/\Delta I],\\ n = 1 \div N_b;\ J_{wnm} = [I_{wnm}/\Delta I],\ n = 1 \div N_w \tag{3}$$

where $[x]$ is the integral part of x value; ΔI is the quantization interval. The quantization interval ΔI is connected to the capacity k of the analog-to-digital coder $\Delta I = I_{lim}/2^k$, where I_{lim} is the maximum possible level of signals received from radiometric detectors. In case of pulsed X-ray sources, images J_w, J are additionally normalized by the signal from the reference channel.

At the next stage, a digital radiographic image Y is composed from J_b, J_w, J images:

$$Y_{nm} = \ln(J_{0m} - \sigma_m) - \ln(J_{nm} - \sigma_m),\ \sigma_m = \sum_{n=1}^{N_b} J_{bnm} \Big/ N_b,\ J_{0m} = \sum_{n=1}^{N_w} J_{wnm} \Big/ N_w. \tag{4}$$

The final digital image Y can be analyzed by an operator or a specialized algorithm oriented to the detection, identification, and estimation of the defect impact degree. In the first case, the image Y is transformed to a digital gray-scale picture Y_{GS}

$$Y_{GS\ nm} = M_{hl}(Y_{nm} - S_b)/(S_w - S_b), \tag{5}$$

where S_b, S_w are respectively black and white levels in Y image; M_{hl} is the number of halftones or gray gradations. The approach, when $S_b = \min_{n,m} Y_{nm}$, $S_w = \max_{n,m} Y_{nm}$ at m=1…M; n=1…N, seems to be more expedient.

The halftone image Y_{GS} is analyzed to detect defects and estimate their validity. The image obtained by the X-ray computed tomography is presented in Figure 2 which illustrates the arrangement of strands in wire rope having the diameter of 25 mm.

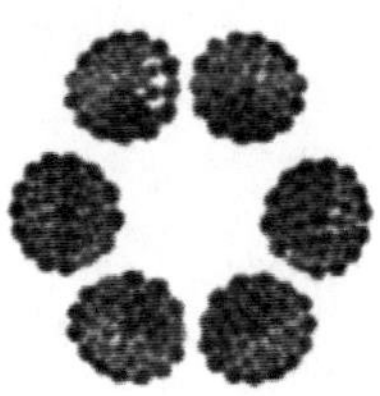

Figure 2. CT-scan image of the wire rope.

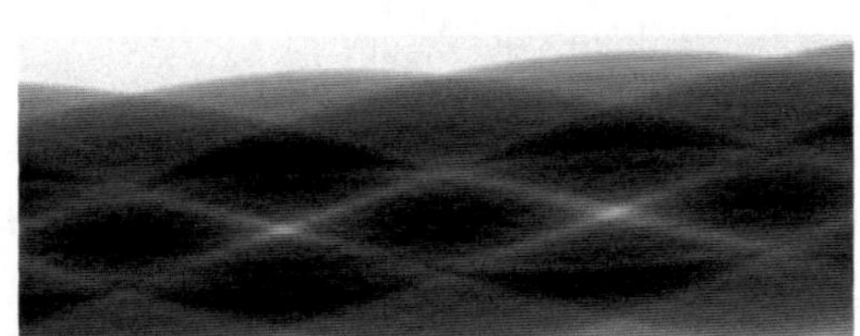

Figure 3. A typical digital radiographic image of wire rope.

The image analysis of wire rope CT-scans is difficult for an operator due to a heterogeneous distribution of intensity in the image. These difficulties can be eliminated by the use of specialized algorithms calculated for processing the radiographic images as described, for example, in the work of Osipov *et al*. [8].

A digital radiographic image of the wire rope with the internal structure shown in Figure 2 is presented in Figure 3. The experiment was carried out on the tomograph with X-ray tube 0.4 mm focal spot, 450 kV voltage; and the PerkinElmer X-ray detector having 200 μm pitch. The best results were obtained at 200 keV radiation energy with the use of filter consisting of a 2 mm thickness sheet of copper and 2.2 projection magnification.

5. X-ray beam hardening correction algorithm

The traditional equation for the radiometric signal I and the thickness ρH (g/cm^2) of the attenuation material is as follows [12]:

$$I/I_0 = e^{-m_{ef}(E_{\max},\rho H)\rho H}, \tag{6}$$

where $m_{ef}(E_{max}, \rho H)$ is the mass attenuation coefficient of photons having energy E_{max}; I_0 is the radiometric signal without a test object.

It is obvious that X-ray beam hardening correction algorithm is reduced to the estimation of ρH parameter. The following main methods are used to calculate the correction algorithm: graphic, inverse functions, analytical approximations, and dual energy X-ray imaging.

$$m_{ef}(E_{max}, \rho H)\rho H = -\ln(I/I_0) = Y = F(\rho H). \tag{7}$$

The graphic method includes the construction of the function $F(\rho H)$ graph by the results of experimental research or mathematical simulation. The function $F(\rho H)$ graph is constructed for the fixed scanning geometry and pre-set E_{max} value. ρH is X-axis and F is Y-axis. The intersection point of the line is $Y=Y_1$, where Y_1 is the measured signal, and the curve $F(\rho H)$ is just the target value of the thickness ρH that corresponds to Y_1 signal.

The method of inverse functions estimate the thickness ρH from

$$\rho H = F^{-1}(Y), \tag{8}$$

where F^{1} is the inverse function. Let parameter ρH takes values of $\rho H_1, \rho H_2, \ldots, \rho H_n$ which correspond to signal values of $Y_1, Y_2, \ldots, Y_n$. In case the measuring signal takes Y value, the value of ρH_i can serve as a crude estimate of ρH target value for which the deviation of $|Y_k - Y|$ at $k=1..n$ is minimum. The more accurate ρH measurement can be provided by the linear interpolation

$$\rho H \approx \rho H_i + (Y - Y_i)(\rho H_{i+1} - \rho H_i)/(Y_{i+1} - Y_i). \tag{9}$$

In case the value n is large and calibration thicknesses $\rho H_1, \rho H_2, \ldots, \rho H_n$ are close to each other, ρH value obtained from equation 9 can be rather exact. Also, it will be more exact if measured by spline interpolation as described by Chen *et al.* [13], the function F^{1} being interpolated. Spline interpolation factors $a_i, b_i, c_i, d_i, i=1\ldots n$ are at stages of calibration measurements.

The analytical approach is based on the description of $Y(\rho H)$ function:

$$Y(\rho H) = \left(1 + A/\left(1 + B\rho H + C(\rho H)^2\right)\right)D\rho H, \tag{10}$$

where A, B, C, D are approximation coefficients depending on $E_{\max}$, effective atomic number of the attenuation material Z, the detector sensitive volume, etc. equation 10 is the approximation refinement as described in works [12, 14]. Coefficients A, B, C, D are ascertained at a stage of calibration. In the course of the formation and transformation of the radiographic image for each of its point, equation 10 can be solved with its left side measured relative to the known ρH value. In order to rapidly solve solutions similar to (10), the abovementioned method of inverse functions can be used.

In terms of wire rope applications, the dual-energy X-ray imaging is considered in detail in works of Dai *et al.* [9] and Osipov *et al* [10]. A method of correction of the factor under analysis is chosen by the rope diameter and the requirements for the detection and identification of its defects as well as their validity. The dual-energy X-ray imaging is the most universal method in this case.

Image Y^* formed at the algorithm output can be obtained from

$$Y^* = \left\{\rho H^*_{nm};\ m=1...M;\ n=1...N\right\}. \tag{11}$$

Further, it is image Y* that is subjected to processing with a view to detect defects of the channels, and carry out their identification, and rationale for validity.

6. The analysis of digital radiographic images of wire ropes

Osipov *et al* [8, 10] suggested the approach to the analysis of digital radiographic image algorithms based on the closeness of certain parameters of all unit cross-sections of the reference wire rope or its sections with the length equaling to the lay. Let these parameters be invariants of the wire rope, and in relation to radiographic images – invariants of radiographic images of wire ropes. Let us give several preliminary apparent notes connected to certain invariants of radiographic images based on Figure 3.

Note 1. There is a symmetric axis of the wire rope shadow image. The presence a symmetric axis generates the existence of a range of invariants of digital radiographic images of wire ropes. Let us introduce the coordinate system *OXY* in the wire rope image. Axis *OX* is the symmetric axis of the shadow rope image. Let us introduce $f_1(x)$ and $f_2(x)$ functions that restrict the wire rope image at the top and at the bottom. Thus, Note 2 can be stated to introduce $f_1(x)$ and $f_2(x)$ functions.

Note 2. Functions $f_1(x)$ and $f_2(x)$ are periodical and connected by $f_1(x) = -f_2(x)$ relation.

Invariants defined by $f_1(x)$ and $f_2(x)$ functions include: maximum transverse size D_{max} of the rope shadow in image; minimum transverse size D_{min} of the rope shadow in image; distance L between locations of the adjacent maxima and minima of these functions.

Strand boundaries are rather clearly seen in Figure 3. The additional invariants are connected with this fact and the significant difference between the maximum brightness of image fragments Y_F. For example, the distance L_w between the white spot centers located on the axis of the shadow image can be chosen as invariant.

One more note is connected with the stability of a certain mass property of the wire rope. It is obvious that the rope sections of the similar length have the similar weight. Let us reformulate this note in terms of the analysis of the final radiographic image of the wire rope.

Note 3. The weight W of the wire rope section limited by the X-ray radiation beam, stays practically unchanged during the formation of the unit line of the final radiographic image.

Let W be the mass of the rope cross-section. The analysis of classical radiographic images or of the dual energy images (Equation 11) allows the accurate measurement of the rope cross-section [15].

The digital radiography provides the solution of such problems as determination of different defects; measurement of defect sizes both in the scanning direction and the direction normal to it; identification of defects. In the work of Osipov *et al* [8], the approaches to the defect identification of wire ropes are considered, namely: broken wire, wear, squashing, local increase of wire rope diameter, corrosion. These approaches are based on the defect formation. Apparently, the identification of certain defect types can be directly connected with the analysis of deviations from reference values (index *nor*, see Table 1) of radiographic image invariants. The Table 1 below presents the more widespread defect types with the abovementioned invariants for wire ropes with fiber cores. Also, visual defect characters and recommendations are given for the use of additional or alternative testing techniques.

7. Conclusion

Equations (1)–(11) sequentially describe distinctive properties of generation and analysis of digital radiographic images of wire ropes allowing to use them for further research that concerns calculations of the image processing algorithm and wire rope proper identification. Results obtained will foster the improvement of NDT techniques in relation to wire rope in-process and in-service testing and the increase of their safety and durability.

Table 1. Invariants of radiographic images and main types of defects.

Main invariants	Defects	Defect visualization
Gradual change of brightness in image fragments	Broken wire	White spot on the background, $Y_F > Y_{nor}$
Maximum transverse size of the rope shadow in the image	Squashing	$D_{max} > D_{nor}$ or $D_{max} < D_{nor}$, but $W \approx W_{nor}$
Mass of the rope cross-section	Wear	$W < W_{nor}$
Maximum transverse size of the rope shadow in the image	Local increase of wire rope diameter	$D_{max} > D_{nor}$, $W \approx W_{nor}$, brightness of the central part is close to the reference, brightness difference increase in image fragments
	Internal corrosion of wire rope	$D_{max} > D_{nor}$, $W \approx W_{nor}$, different brightness in image fragments. Dual energy X-ray imaging is recommended
	Local diameter reduction due to a fiber core	$D_{max} < D_{nor}$, $W \approx W_{nor}$
	Foreign inclusions	$D_{max} > D_{nor}$ and $W > W_{nor}$
Image symmetry	Wire rope rupture	Asymmetric distortion or formation of darker areas in the image. Several camera angles are recommended
Brightness of image fragments	Separate strand or wire pushing out	$D_{max} > D_{nor}$. Several camera angles are recommended
Distance between centres of white spots located on the axis of the shadow image	Wire rope extension	$L_w > L_{w\ nor}$ and $L > L_{nor}$, $\Sigma W = \Sigma W_{nor}$

Acknowledgements
The authors acknowledge the financial support of the Ministry of Education and Science of the Russian Federation for carrying out this research.

References

[1] Zhang D *et al* 2013 *J. Nondestruct. Eval.* **32** 37–43
[2] Krešák J *et al* 2012 *Metalurgija-Zagreb* **51** 485–488
[3] Hua-Ming L *and* Yi-Mei M 2013 *Rus. J. Nondestruct. Testing* **49** 602–609
[4] Liu H P, Miao C Y *and* Li X G 2013 *Adv. Mat. Res.* **722** 572–575
[5] Xian-Guo L *et al* 2012 *Int. J. of Digital Content Technology and its Applications* **6** 226–234
[6] Ewert U *et al* 2010 *Mater. Eval.* **68** 163–168
[7] Ewert U *et al* 2012 *Mater. Eval.* **70** 955–964
[8] Osipov S P *et al* 2013 *Tekhnologii Tekhnosfernoi Bezopasnosti* **51**
[9] Dai Y, Miao C *and* Rong F F 2009 *Int. Symp. Photoelectronic Detection and Imaging: International Society for Optics and Photonics* **7385**
[10] Osipov S P *et al* 2012 *Tekhnologii Tekhnosfernoi Bezopasnosti* **44**
[11] Sidulenko O A *et al* 2007 *Instruments and Systems: Monitoring, Control, and Diagnostics* **3** 62–64
[12] Nedavnii O I *and* Osipov S P 1994 *Rus. J. Nondestruct. Testing* **9** 92–95
[13] Chen F *and* Wong R J Y 2014 *J. Comput. Appl. Math.* **255** 282–296
[14] Pease B J *et al* 2012 *Constr. Build. Mater.* **36** 419–429
[15] Osipov S P *et al* 2015 *Adv. Mat. Res.* **1085** 455-459

IOP Conf. Series: Materials Science and Engineering **132** (2016) 012009 doi:10.1088/1757-899X/132/1/012009

Regimes of Decomposition of Clathrate in Natural Strata Purged by Methane

M Khasanov[1] and V Shagapov[2]
[1]Associate Professor, Sterlitamak Branch of Bashkir State University, Sterlitamak, Russia
[2]Principal researcher, IME KazSC RAS, Kazan, Russia

E-mail: hasanovmk@mail.ru

Abstract. The process of decomposition of a methane clathrate in a finite-length stratum initially saturated with methane clathrate and methane, which is purged by warm methane, is studied. The influence of the initial parameters of the stratum and purging conditions on the evolution of methane clathrate temperature and saturation is examined. The existence of solutions is demonstrated, which predict methane clathrate decomposition both on the frontal surface and in the volume zone.

1. Introduction

In the current paper we consider the features of methane clathrate decomposition at warm methane injection into a stratum of a finite length, saturated initially with methane clathrate and methane. The self-similar solution to the problem on cold methane injection into a stratum, saturated with water and methane, was obtained in [1–4].

2. Basic Equations

Let's consider a stratum, saturated initially with methane clathrate and methane, whose temperature T_0 and pressure p_0 meet the condition of stable existence of methane clathrate and methane. Let injection of methane with pressure p_w and temperature T_w $(T_w > T_0)$ is started through the left surface of a stratum $(x = 0)$. These parameters meet the conditions of joint existence of water and methane, and they are kept constant during the entire process of methane clathrate decomposition into the water and gas. We will assume that there are three zones formed in the porous medium, where water, methane clathrate and methane are in different states. Hence, the pores of the first (nearest) zone are filed with water and methane, and the pores of the third (distant) zone are filed with methane clathrate and methane. The second (transition) zone contains methane clathrate and products of its decomposition (water and methane) in the state of phase equilibrium. At that, there are two moving surfaces $x = x_{(i)}$ $(i = n, d)$, which separate the said zones: surface $x = x_{(d)}$ separates the third and transition zones and surface $x = x_{(n)}$ separates the first and transition zones.

To describe the processes of filtration and heat transfer at methane injection into the porous medium, accompanied by methane clathrate decomposition, let's assume the unitemperature model of stratum, i.e., the temperature of contained substance (water, methane, and methane clathrate) and temperature of the stratum coincide at every point. Let's also assume that bed porosity is constant; its water, skeleton of the porous medium, and methane clathrate are incompressible, and methane is calorically perfect gas:

$$m,\ \rho_{sk},\ \rho_h,\ \rho_l = \text{const},\quad p = \rho_g R_g T. \tag{1}$$

Here m is porosity of the stratum, ρ_j ($j = g$, h, sk, and l) are true phase densities (subscripts g, h, sk, l are attributed to the parameters of gas, methane clathrate, skeleton of the porous medium, and water, respectively), T is temperature of the stratum, p is pressure of the stratum, R_g is reduced methane constant. Methane clathrate is the two-species system with mass concentration of methane G.

The conditions of mass conservation for methane and water in the one dimensional statement can be presented as [1, 2]:

$$\frac{\partial}{\partial t}\left(mS_g\rho_g\right)+\frac{\partial}{\partial x}\left(mS_g\upsilon_g\rho_g\right)=-mG\rho_h\frac{\partial S_h}{\partial t}, \tag{2}$$

$$\frac{\partial}{\partial t}\left(mS_l\right)=-m\left(1-G\right)\frac{\rho_h}{\rho_l}\frac{\partial S_h}{\partial t}, \tag{3}$$

$$S_g+S_h+S_l=1,$$

where υ_g is gas phase velocity, t is time, S_j is saturation of pores by the jth phase.

For methane filtration lets' assume Darcy law:

$$mS_g\upsilon_g=-\frac{k_g}{\mu_g}\frac{\partial p}{\partial x}. \tag{4}$$

where μ_g is dynamic viscosity methane, k_g is gas phase permeability:

$$k_g=k_*\frac{\left(mS_g\right)^3}{\left(1-mS_g\right)^2}\approx k_0S_g^3,\ \left(k_0=k_*m^3\right). \tag{5}$$

Here k_0 is absolute permeability of the porous medium.

The energy equation can be written as:

$$\rho c\frac{\partial T}{\partial t}+c_g\rho_gmS_g\upsilon_g\frac{\partial T}{\partial x}=\frac{\partial}{\partial x}\left(\lambda\frac{\partial T}{\partial x}\right)+m\rho_hL_h\frac{\partial S_h}{\partial t}, \tag{6}$$

$$\rho c=\left(1-m\right)\rho_{sk}c_{sk}+m\sum_{j=g,l,h}S_j\rho_jc_j,$$

$$\lambda=\left(1-m\right)\lambda_{sk}+m\sum_{j=g,l,h}S_j\lambda_j,$$

where T is temperature of the stratum, λ_j ($j = sk$, g, h, l) and c_j are coefficient of heat conductivity and specific heat capacity of skeleton, methane, methane clathrate and water, values λ and ρc are coefficient of heat conductivity and specific volumetric heat capacity of "porous stratum - methane clathrate" system, and L_h is specific heat of methane clathrate decomposition into the water and gas.

The regions, where methane, water and methane clathrate can be in different states, are formed in the stratum at methane clathrate decomposition. On the surfaces between these regions, where phase saturations, heat and mass flows undergo jumps, the relationships following from the equations of energy balance and mass balance are satisfied [5]:

$$\left[m\left(\rho_lS_l+\rho_hS_h\left(1-G\right)\right)\dot{x}_{(i)}\right]=0,\quad \left[m\left(\rho_gS_g\left(\upsilon_g-\dot{x}_{(i)}\right)-\rho_hS_hG\dot{x}_{(i)}\right)\right]=0, \tag{7}$$

$$\left[\lambda\frac{\partial T}{\partial x}\right]=\left[m\rho_hL_hS_h\dot{x}_{(i)}\right].$$

Here $[\psi]=\psi^{+}-\psi^{-}$ is the surge of parameter ψ on surface $x=x_{(i)}$, $\dot{x}_{(i)}$ is velocity of this surface motion, subscripts $i=d, n$ correspond to values on the distant and nearest surfaces, respectively. The pressure p and temperature T on both surfaces are assumed to be continuous.

In the second zone containing water, methane clathrate and methane where methane clathrate decomposition occurs, the equation of phase equilibrium is accepted [6]:

$$T=T_0+T_*\ln\left(\frac{p}{p_{s0}}\right), \tag{8}$$

where T_0 is temperature of the porous medium in the initial state, p_{s0} is equilibrium pressure, corresponding to the temperature T_0, and T_* is empirical parameter that depends on the form of clathrate.

Let the porous medium is initially saturated with methane clathrate and methane, whose temperature and pressure correspond to conditions of stable existence of methane clathrate and methane and they are the same in the whole porous medium:

$$p=p_0,\ T=T_0,\ S_h=\nu\ \ (0\le x\le L,\ t=0). \tag{9}$$

Injection of warm methane is started through the left surface of the porous medium under temperature T_w and pressure p_w, its values meet the conditions of joint existence of water and methane:

$$p=p_w,\quad T=T_w\quad (x=0,\ t>0). \tag{10}$$

On the right surface of the porous medium, we impose conditions modeling the absence of the heat flow and constant pressure:

$$p=p_0,\quad \frac{\partial T}{\partial x}=0\quad (x=L,\ t>0). \tag{11}$$

Applying the equation of the methane state and Darcy law to relations (2) and (6) and assuming that the characteristic temperature drops in the porous stratum are insignificant, we can get the equations of thermal conductivity and piezoconductivity, total for all porous medium regions:

$$\frac{\partial p}{\partial t}=\frac{k_g}{m\mu_g S_g}\frac{\partial}{\partial x}\left(p\frac{\partial p}{\partial x}\right)-\Re T\frac{\partial S_h}{\partial t}, \tag{12}$$

$$\frac{\partial T}{\partial t}=\frac{\partial}{\partial x}\left(\kappa^{(T)}\frac{\partial T}{\partial x}\right)+\frac{c_g k_g}{\mu_g\rho c}\frac{p}{R_g T}\frac{\partial p}{\partial x}\frac{\partial T}{\partial x}+\Delta T\frac{\partial S_h}{\partial t}, \tag{13}$$

where $\Re=\dfrac{G\rho_h R_g}{1-S_{l0}}$, $\Delta T=\dfrac{m\rho_h L_h}{\rho c}$, $\kappa^{(T)}=\lambda/\rho c$.

In the transition zone, where the temperature T and pressure p of the stratum are connected by the conditions (8):

$$\frac{\partial T}{\partial x}=\frac{T_*}{p}\frac{\partial p}{\partial x},\quad \frac{\partial^2 T}{\partial x^2}=\frac{T_*}{p}\frac{\partial^2 p}{\partial x^2}-\frac{T_*}{p^2}\left(\frac{\partial p}{\partial x}\right)^2.$$

Using the last relationships, from the conditions (12) and (13) we obtain the equations describing distributions of methane clathrate saturation and pressure in the transition zone of the porous medium:

$$\frac{\partial p}{\partial t}=\frac{k_g}{m\mu_g S_g}\frac{\partial}{\partial x}\left(p\frac{\partial p}{\partial x}\right)-\frac{\Re T}{\Delta T}\frac{\partial^2 T}{\partial x^2}+\frac{c_g k_g \Re}{\mu_g \rho c}\frac{p}{\Delta T\cdot R_g}\frac{\partial p}{\partial x}\frac{\partial T}{\partial x}, \tag{14}$$

$$\frac{\partial S_h}{\partial t}=\frac{1}{\Delta T}\left[\frac{\partial T}{\partial t}-\frac{\partial}{\partial x}\left(\kappa^{(T)}\frac{\partial T}{\partial x}\right)-\frac{c_g k_g}{\mu_g \rho c}\frac{p}{R_g T}\frac{\partial p}{\partial x}\frac{\partial T}{\partial x}\right]. \tag{15}$$

Since, methane clathrate decomposition occurs within the transition zone, we will assume that on the nearest surface of clathrate decomposition $x = x_{(n)}$ there is the surge of methane clathrate saturation from $S^-_{h(n)}=0$ to $S^+_{h(n)}>0$. Using relationships (7), we obtain the balance equations the energy and methane mass:

$$\frac{k^+_{g(n)}}{\mu_g}\left(\frac{\partial p}{\partial x}\right)^+_{(n)}-\frac{k^+_{g(n)}}{\mu_g}\left(\frac{\partial p}{\partial x}\right)^-_{(n)}=m\dot{x}_{(n)}\left[S^-_{g(n)}-S^+_{g(n)}-\frac{\rho_h G S^+_{h(n)}}{\rho_{g(n)}}\right], \tag{16}$$

$$\lambda\left(\frac{\partial T}{\partial x}\right)^+_{(n)}-\lambda\left(\frac{\partial T}{\partial x}\right)^-_{(n)}=m\rho_h L_h S^+_{h(n)}\dot{x}_{(n)}.$$

Hereinafter the subscripts "plus" and "minus" are attributed to the values before and after the surface, respectively.

On the distant surface of clathrate decomposition $x = x_{(d)}$ the value of methane clathrate saturation S_h is assumed continuous and equal to initial methane clathrate saturation of the porous medium. Then, on the basis of equations (7) the balance equations for the heat and methane mass on this surface take form:

$$\left(\frac{\partial p}{\partial x}\right)^-_{(d)}=\left(\frac{\partial p}{\partial x}\right)^+_{(d)},\quad \left(\frac{\partial T}{\partial x}\right)^-_{(d)}=\left(\frac{\partial T}{\partial x}\right)^+_{(d)}. \tag{17}$$

The method of front capture by nodes of the spatial grid was used for solution to problem (12)–(15) with initial and surface Equations (9)–(11) and equations on moving surfaces of clathrate decomposition (16) and (17).

3. Results

For parameters describing the "porous stratum - methane clathrate" system, the following values were adopted: L = 1 m, $k_0 = 10^{-14}$ m^2, ν = 0.2, T_0 = 280 K, p_0 = 4 MPa, T_w = 290 K, $p_w = 7\cdot 10^6$ Pa, G = 0.13, $p_{s0} = 5.45\cdot 10^6$ Pa, T_* = 10 K, R_g = 530 J/(K·kg), c_g = 1560 J/(K·kg), $\rho_h = 900$ kg/m^3, $\lambda_h = 2.2$ W/(m·K), c_h = 2000 J/(K·kg), $\rho_l = 1000$ kg/m^3, c_l=4000 J/(K·kg), $\lambda_l = 0.6$ W/(m·K), $\rho_{sk} = 2.1\cdot 10^3$ kg/m^3, c_{sk} = 900 J/(K·kg), $\mu_g = 10^{-5}$ kg/(m·sec), $\lambda = 2$ W/(m·K), $\rho c = 2.5\cdot 10^6$ J/(K·m^3), $L_h = 4.5\cdot 10^5$ J/kg [7–8].

Distributions of methane clathrate saturation and temperature of the system (solid lines) are shown in figure 1. The initial value of pressure in the reservoir was taken $p_0 = 6\cdot 10^6$ Pa. The dashed line is the equilibrium temperature of the stratum. According to the figure, at above listed values temperature and pressure of injected methane clathrate decomposition until $t \approx 4\cdot 10^4$ sec occurs on the frontal boundary.

However, formation of the volume zone of methane clathrate decomposition is observed then. This is due to the fact that at injection there is a heat flow from the left surface deep into the porous medium, which causes a temperature increase in the zone saturated with methane clathrate and methane and formation of the volume zone of methane clathrate decomposition. At that, two moving surfaces are formed; one of these surfaces by moment $t \approx 5 \cdot 10^4$ sec reaches the right surface of the porous medium. Then, two zones exist in the porous medium: the first (nearest) one saturated with water and methane and the distant (second) one containing methane clathrate and products of its decomposition (water and methane); they are separated by the only moving surface. Thus, at methane purging of the porous medium clathrate decomposition occurs firstly on the frontal boundary and then, in the volume zone of methane clathrate decomposition.

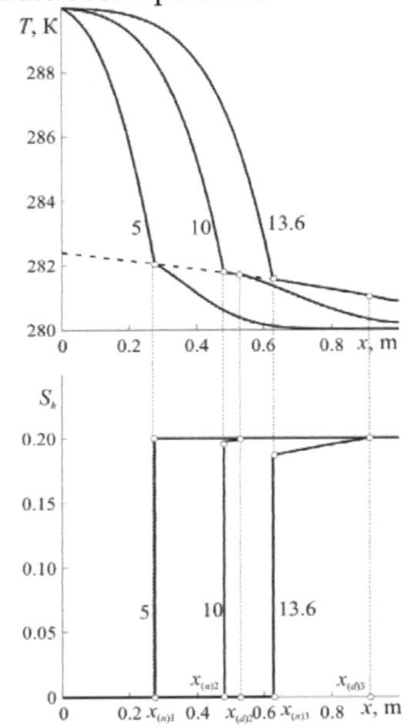

Figure 1. Distributions of methane clathrate saturation and temperature (numbers on curves – time in hours).

4. Conclusion

The process of methane clathrate decomposition in a porous medium, initially saturated with methane and methane clathrate, was studied for the case of warm methane purging of the stratum. It is shown that methane clathrate decomposition occurs firstly on the frontal boundary and then in the volume zone.

Acknowledgement
This work is supported by the Russian Science Foundation under grant 15-11-20022.

References

[1] Shagapov V Sh *et al* 2015 *Int. J. of Heat and Mass Transfer* **84** 1030–1039
[2] Khasanov M K 2015 *Thermophysics and Aeromechanics* **22** 255–265
[3] Pritulov A M *et al* 2007 *Russian Physics Journal* **50 (2)** 187–192 DOI: 10.1007/s11182-007-0026-3
[4] Surzhikov A P *et al* 2006 *Russian Physics Journal* **49 (5)** 506–510 DOI: 10.1007/s11182-006-0133-6
[5] Nigmatulin R I 1991 *Dynamics of Multiphase Media* (New York: Hemisphere Publ. Corp.) **1**
[6] Khasanov M K *et al* 2015 *J. of Engineering Physics and Thermophysics* **88** 1052–1061
[7] Sloan E D and Koh C A 2008 *Clathrate Hydrates of Natural Gases* (BocaRaton: CRCPress)
[8] Sloan E D 1998 *Natural Gas Clathrate Hydrates* (New York: Marcel Dekker)

IOP Conf. Series: Materials Science and Engineering 132 (2016) 012010 doi:10.1088/1757-899X/132/1/012010

Development and Experimental Investigations of Motion Detection Module for Smart Lighting System

I Matveev[1], E Siemens[2, 3] A Yurchenko[2] and D Kuznetsov[4]
[1]Student, National Research Tomsk Polytechnic University, Tomsk, Russia
[2]Professor, National Research Tomsk Polytechnic University, Tomsk, Russia
[3]Senior Researcher, National Research Tomsk State University, Tomsk, Russia
[4]Postgraduate Student, National Research Tomsk State University, Tomsk, Russia

E-mail: niipp@inbox.ru

Abstract. This work considers motion sensors as parts of the smart lighting system on basis of Beaglebone microcomputer. Detection system is designed for the smart lighting system. Experimental investigations of the detection system were made with different motion sensors. Based on the results comparative analysis was performed and optimal conditions for the detection system operation were found.

1. Introduction

Modern city lightning systems have raised requirements to energy conservation and resource efficiency due to growing power consumption for city lightening. Smart lightening systems meet such requirements.

Advantages of the smart lightening system over ordinary one:

– standard lightning systems have preset lightning time which does not depend on actual daylight hours and weather conditions that is solved by the smart lightning system that uses this dependence and assumes necessary measures;

– switching universally used lightning systems on and off do not depend on the presence of pedestrians and moving vehicles resulting in pointless energy consumption during long period of time [1–5];

– smart lightning control systems in contrast with ordinary ones monitor current status of each LED lamp and transfer information to the server that provides more efficient operation and timely failure removal;

– hardware could be easily updated in accordance with installation location of the smart lightning system — busy highways, quiet suburban streets or city part alleys;

– as smart lightning systems presuppose continuous monitoring and prompt troubleshooting that provides constant streets lighting and consequently influences on road safety [6].

To control smart lightning systems there are integrated systems that control lightning level, process sensor data signals and control communication with each other. Type of used sensors determines composition and functioning of the whole system as a sensor data signal ultimately indicates about the need to adjust lightning that is performed by the feedback of the majority of system units.

2. Smart lightning system structure

Lightning system is a combination of a central device and connected to it street LED lamps (figure 1).

Beaglebone microcomputer acts as a central control device. Each lamp applies to communication and detection system. First one serves to connect a lamp with Beaglebone, the second – to detect a monitored object (a pedestrian/moving vehicle passing by). When a pedestrian enters the monitored area and consequently when the sensor detects him/her a signal comes to Beaglebone that in turn switches the most neighbouring to the pedestrian LED lamps.

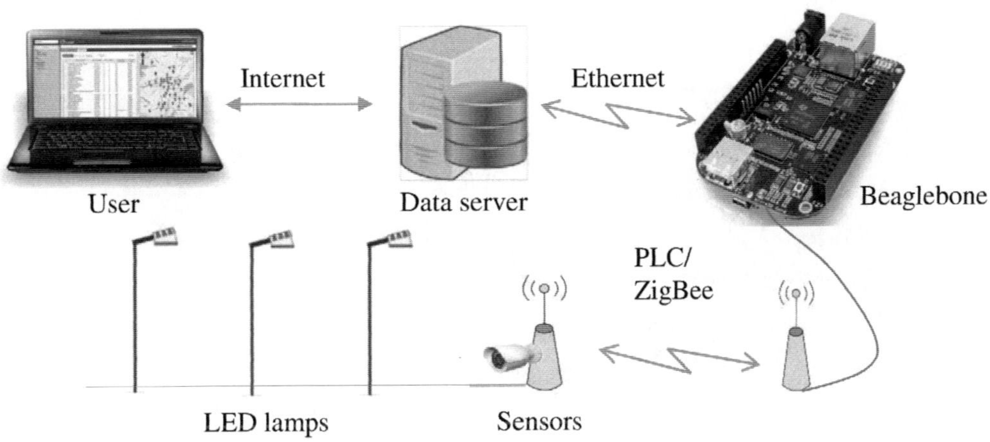

Figure 1. Smart lightning system structure.

Software components are used in accordance with hardware. Developed program runs on Beaglebone that sends appropriate control signals to lamps. The complexity of it consists in creation of a reliable method of this or that object sensing. Such challenge is caused by the level of the design of motion sensors used in lightning systems. Currently they are often activated falsely or are not activated wrongly and also go off when the monitored object stops. Solution for this problem may be found with combined sensors and perfect algorithms for processing data.

Communication between LED lamps and microcomputer is provided by the power line communication (PLC) technology or ZigBee wireless technology.

PLC communication center should be capable to transfer data on LED condition, voltage, current strength, consumed power, capacity with transmission facilities using protocol ANSI/EIA 709.1 [7 – 8].

In comparison with PLC ZigBee protocol allows to avoid pauses in operation and fails caused by noise and signal attenuation using wireless signal transmission. This protocol is based on standard IEEE 802.15.4 and has following features: low price, low power consumption, high reliability [9].

All data on events taking place in the lightning system after processing come to the database server and are delivered to a dispatcher.

3. Experimental unit

The most widespread motion sensors used in smart lightning systems are radio-wave, infrared, ultrasonic and combined.

An experimental unit was assembled. It included detection system of the smart lightning system (figure 2). Its main components are BeagleBone Black microcomputer, infrared sensor DP104, ultrasonic sensor SRF08, combined sensor Sockol-3, power supply unit with outputs 5V, 12V.

Microcomputer BeagleBone Black. Microcomputer BeagleBone Black is used as a control unit that performs a program of information readout from sensors. BeagleBone Black specifications are presented in table 1 [10].

Infrared sensor DP104. Specifications of DP104 infrared sensor are presented in table 2.

Ultrasonic sensor SRF08. SRF08 Ultrasonic sensor connects BeagleBone Black microcomputer with a serial data bus I^2C.

In the beginning microcomputer initializes address of the sensor connected through the protocol I^2C [10]. To scan the monitored area with the sensor requires a command that sends certain value to the command register SRF08. The next step is reading out data from sensor registers. Data are received in hexadecimal format and carry information about distance to the object in cm. Features of SRF08 ultrasonic sensor are presented in table 2.

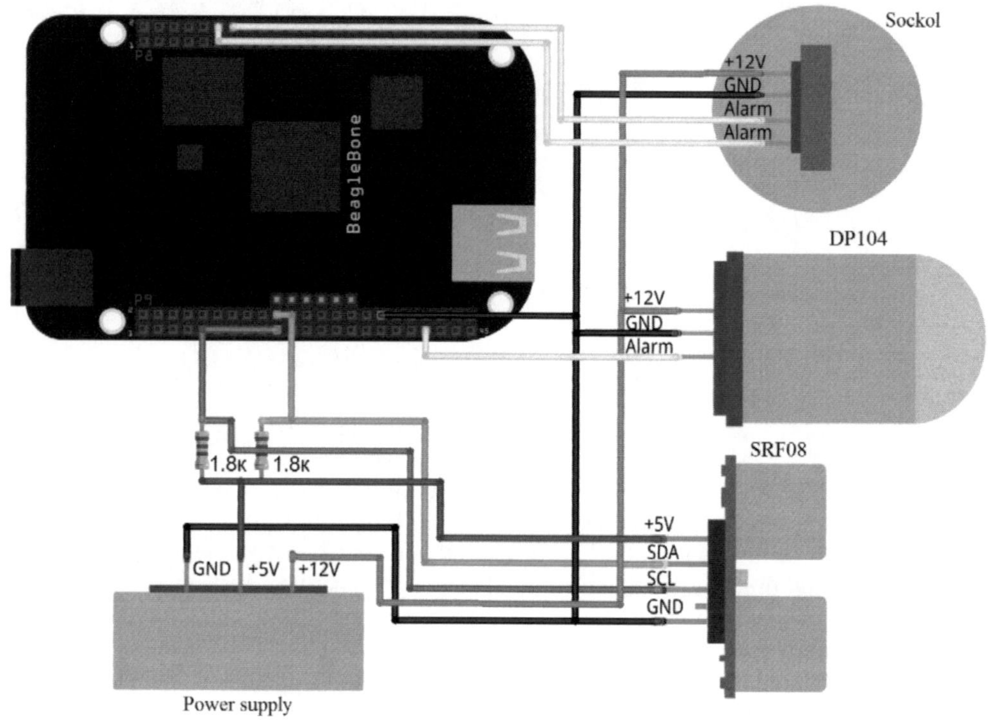

Figure 2. Experimental mock-up.

Table 1. Specifications of BeagleBone Black.

Processor	1 GHz AM335x
Memory	512 Mb DDR3 (faster 1.6 Gb/s and lower power)
Storage	On-board 2 Gb eMMC (4 Gb eMMC on the Revision C board) and micro-SD card slot
Video	On-board HDMI
Debugging	JTAG header present but not populated
Serial Connection	TTL header present but separate cable needed
Input/Output Headers	GPIO, analog outputs/inputs, 5 V, 3.3 V, 1.8 V power supply, timers, buses: I^2C, UART, CAN, SPI, GPMC, MMC, LCD, McASP

Combined sensor Sockol-3. When a person enters the detection area both detection channels go off (in any consequence). At the same time an alert is produced by breaking output relay contact.

Table 2. Specifications of sensors.

Type of sensor used	DP104	SRF08	Sockol-3
Voltage	DC 12 V	5V	10..15V
Current	-	15mA Typ. 3mA Standby	30mA
Working range	6-10 m	3 cm..6 m	5…10m
Working temperature range	-	-	-30..+50°C
Additional information	time setting: 5s~6min, lux setting: 2~2000 lux	connection: standard I^2C bus	Microwave, infrared types of sensors combined

Sensor continually monitors interference signaling environment in protected room: if interference level through one of channels exceeds tolerable, sensor automatically changes the detection algorithm. The sensor operates during 2 minutes subordinate to the changed algorithm, in 2 minutes it returns to the basic detection algorithm and the cycle begins to repeat. Transfer to another algorithm is indicated by the green light. If operating sensor receives an alert by the detection algorithm in conditions of increased interference, immediately after its termination the sensor returns to the basic detection algorithm.

Specifications of Sockol-3 combined sensor are presented in table 2.

4. Experimental results and analysis

Main aim of experimental investigations of motion sensors is comparative analysis of existed sensors, determination of optimal conditions for the use in the smart lightning system on the basis of BeagleBone Black microcomputer.

C++ based program was designed. It allows receiving information about sensor actuation at a motion in the sensors' control area.

Examination of sensor sampling results in choosing typical devices from each category that were previously presented and used in the experimental unit. For motion sensors of different categories a chart is built to show areas of their coverage determined by an experiment (figure 3). An experiment was performed in the area covered by all involved sensors. Its aim is to detect response of each sensor to pedestrian's motions.

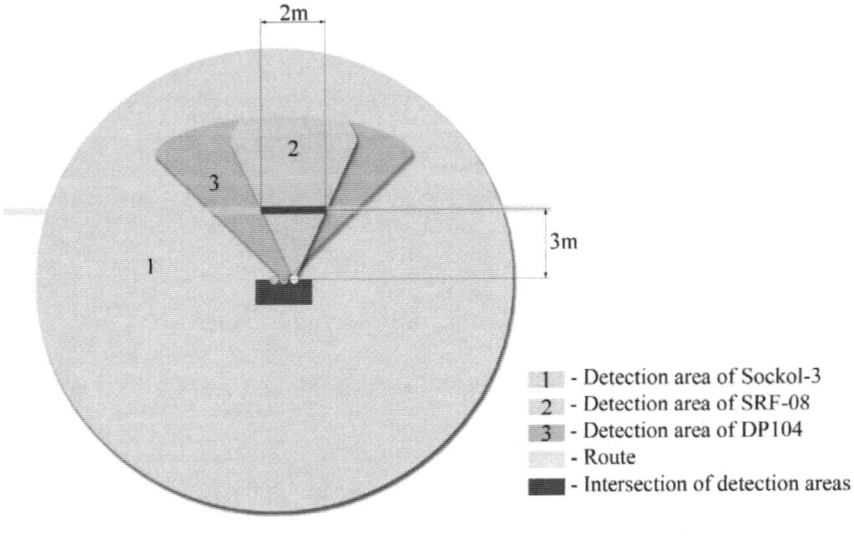

Figure 3. Detection areas of motion sensors and experimental conditions.

Detection of sensor response to pedestrian's motions was in 3 meters from sensors' location. Measurement data are presented on figure 4. Pedestrian motion distance in the area of all sensor intersection is 2 meters, at that distance for Sockol-3 and DP104 is two times longer.

In the course of measurements it was determined that the fastest response time appertains to ultrasonic sensor SRF08 that is second splits. Delay when a pedestrian enters the controlled area is typical for a infrared sensor and a combined one. It is 0.5 second. At that a delay after pedestrian leaving the controlled area is 5 seconds for DP104 and 0.5 second for Sockol-3.

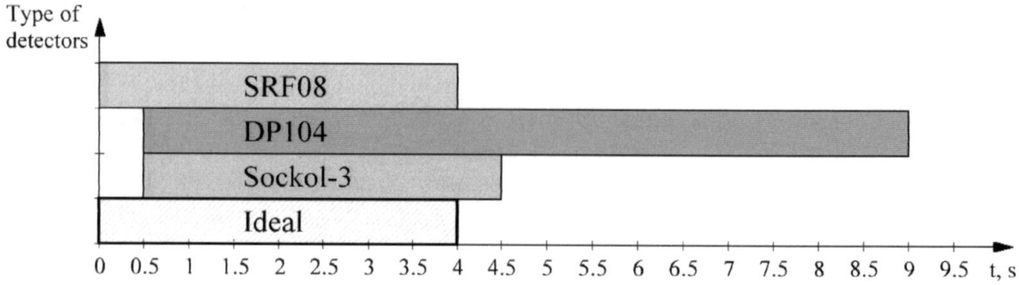

Figure 4. Detection areas of motion sensors and experimental conditions.

According to the results presented on Figures 3 and 4 we can conclude that combined type of sensors is the most optimal option for the use in smart lightning systems. The use of ultrasonic sensor in such systems is hardly possible due to the small angle of sensor action. More than that, this type of sensors has a high price in comparison with others. Also it is possible to use an infrared sensor in smart lightning systems as it has a great angle of action; however it is characterized by a bigger delay in response time.

5. Conclusion

In the course of made investigations different types of motion sensors were analyzed. The most optimal pair of sensors was determined for the use in the smart lightning system. It is realized in the examined combined sensor. The use of combination of detection methods allows increasing accuracy and sensitivity of the lightning system operation.

Diagrams and algorithms of sensor connection to the Beaglebone Black microcomputer were worked out. They have become a basis for an experimental mock-up. Investigations of different sensor types' sensitivity were made. A chart showing coverage areas of examined sensors was built.

Results of this investigation are planned to use in creation of pedestrian motion algorithms in areas of smart lightning systems installation in Tomsk. More than that, the system of communication between LED lamps and control units in the smart lightning systems will be worked out on the basis of the received data.

References

[1] Srivatsa D K *et al* 2013 *Proc. of Texas Instruments India Educators' Conf. (Bangalore)* 103–106
[2] Petrova A Y *et al* 2015 *Technical Physics* **60 (4)** 592–594
[3] Payuk L A *et al* 2016 *Journal of Physics: Conference Series* **671 (1)** 012044 DOI:10.1088/1742-6596/671/1/012044
[4] Surzhikov A.P *et al* 2010 *Journal of Thermal Analysis and Calorimetry* **102 (3)** 883–887
[5] Surzhikov A P *et al* 2012 *Journal of Thermal Analysis and Calorimetry* **108 (3)** 1207–1212

[6] Yoshiura N *et al* 2013 *Proc. of 13th Int. Symp. on Communications and Information Technologies (Samui)* 633–637
[7] ANSI/EIA-709.1-A 1998 *Control Network Protocol Specification*
[8] Cho S K *and* Dhingra V 2008 *Proc. of Int. Symp. Power Line Communications and Its Applications (Seoul)* 396–398
[9] Kaleem Z *et al* 2014 *Proc. of 12th International Conference on Frontiers of Information Technology (Islamabad)* 361–365
[10] Molloy D 2015 *Tools and Techniques for Building with Embedded Linux* (New York: Wiley)

Integrated Method – the Optimum Way to Improve the Quality of Frequency Response Characteristics of the Space Vehicle Attitude Control System

Yu. Britova[1, a], V. Dmitriev[1, a] and T. Kostyuchenko[1, a]

[1]National Research Tomsk Polytechnic University, Russia, Tomsk
[a]britova@tpu.ru

Keywords: Reaction wheel motor, Modal analysis, Vibration characteristics, Ball bearing, Critical angular velocity, 3D models.

Abstract. The integrated method applied to the design of technical systems is a process in which various project, calculation and verification procedures are interconnected and interrelated. The results of procedures are used in a certain sequence, thus ensuring maximum reachable optimality of the system being designed.

Introduction

Executive element of the attitude control system is one of the space vehicle's technical systems. At the moment one of the most popular type of executive elements is the one based on the torque-controlled reaction wheel motor (RWM). Generally RWM (Fig. 1) comprises a solid rotor with expressed rotor rim *1*, fitted onto the supports *2*, an electric motor *3* with rotor attached to the reaction wheel and stator attached to the to the base *4* [1].

Methods. Experimental Procedure

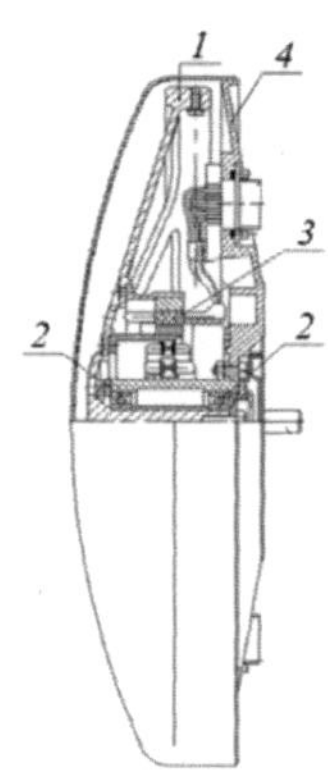

Figure 1. RWM design

Design of RWM, as with any other electromechanical system, starts with calculation of major operation parameters and, subsequently, of required mass-dimensional characteristics. RWMs are designed based on the classical mathematical model shown in the form of Equations (1)–(11). They demonstrate interrelations of its operating parameters.

Due to toughening of the requirements imposed on executive elements of attitude control systems, the mathematical model was subjected to expansions with regard to critical angular velocity, induced perturbated oscillations, resonant angular velocity of the wheel and self frequencies of the system parts.

$H=\frac{1}{32}\pi\gamma h\Omega D_{w}^{4}\left(1+k^{4}\right);$	kinetic momentum	(1)
$m=\frac{\gamma h\left(R^{2}-r^{2}\right)}{2};$	wheel mass	(2)
$N=\frac{1}{\Omega}\left(\frac{C}{Q_{l}}\right)^{3,3};$	operation life	(3)
$P=1.028M_{c}\Omega 10^{5}\eta;$	power consumption	(4)
$M_{r}=\left(M_{c}+M_{bb}+M_{a}\right);$	resistive torque	(5)

$$M_c = \frac{dH}{dt} = J\Omega$$ wheel control torque (6)

$$T = 4T_{const}$$ run-up time (7)

$$M_{bb} = 1.11K\left(1 + \frac{D_{inner}}{d_r}Q_l\right);$$ frictional torque of ball bearings (8)

$$M_a = 2\pi\rho C_{L(T)}\Omega^2;$$ aerodynamic resistance torque (9)

$$C_L = 0.53R_e^{-0.5};$$ (10)

$$C_T = 0.0287R_e^{-0.2};$$ (11)

aerodynamic coefficients for laminar and turbulent flows (10), (11)

$$\omega_{cr} = \frac{1}{\sqrt{0.5(m\alpha - J_Z\beta) + \sqrt{0.25(m\alpha - J_Z\beta)^2 + mJ_Z(\alpha\beta - \xi^2)}}}$$

with

$$\alpha = \frac{ab^2}{3EI_1} + \frac{b^3}{3EI_2},\ \beta = \frac{a}{3EI_1} + \frac{b}{EI_2},\ \xi = \frac{ab}{3EI_1} + \frac{b^2}{2EI_2}$$

critical angular velocity of WM (12)

$$\vartheta = \frac{\left[Gl - (J_Z + J_X)\Omega^2\right](J_Z - J_E)\varepsilon\Omega^2}{(Gl - J_Y\Omega^2)(Gl - J_X\Omega^2) - (J_Z\Omega^2)^2}\sin\Omega t$$ (13)

$$\psi = -\frac{\left[Gl - (J_Z + J_Y)\Omega^2\right](J_Z - J_E)\varepsilon\Omega^2}{(Gl - J_Y\Omega^2)(Gl - J_X\Omega^2) - (J_Z\Omega^2)^2}\cos\Omega t$$ (14)

induced oscillations occurring along two orthogonal axes of WM (13), (14)

$$\omega_r = \pm\sqrt{\frac{Gl(J_X + J_Y) \pm \left[Gl(J_X + J_Y) - \frac{GlJ_XJ_Y}{J_X + J_Y}(1 - \mu^2)\right]}{2J_XJ_Y(1 - \mu^2)}}$$ resonant angular velocity of the wheel (15)

$$f_m = 0.5\left[1 - \left(\frac{2d_{ball}}{D + d}\right)cos\,q\right]\Omega$$ vibratory frequency due to misbalanced ball bearing cage (16)

$$f_{inner} = 0.5\left[1 + \left(\frac{2d_{ball}}{D + d}\right)cos\,q\right]z\Omega$$ vibratory frequency due to defective inner ring of the ball bearing (17)

$$f_{outer} = 0.5\left[1 - \left(\frac{2d_{ball}}{D + d}\right)cos\,q\right]z\Omega$$ vibratory frequency due to defective outer ring of the ball bearing (18)

$$f_b = \left(\frac{d + D}{4d_{ball}}\right)\left[1 - \frac{4d^2_{ball}}{(D + d)^2}cos^2\,q\right]\Omega$$ vibratory frequency due to defective rolling element of the ball bearing (19)

$$f_0 = \frac{d + D}{2d_{ball}}z\Omega$$ vibratory frequency due to rough or oval-shaped rolling elements of the ball bearing (20)

$$f_Z = 0.5z\Omega$$ vibratory frequency due to play in the ball bearing (21)

with R for external radius of the wheel; r for internal radius of the wheel; $k=R/r$; m for RWM weight; γ for specific density of the wheel material, h for rim height; D_w for wheel diameter; Ω for angular velocity of rotation; C for ball bearings performance factor; Q_l for load; η for efficiency; T_m for time constant; K for coefficient of rolling friction; D_{inner} for inner diameter of the ball bearing outer ring; ρ for density of the media around the wheel; R_e for Reynold's number; D_0 for diameter through the centers of the ball bearing rolling elements; z for quantity of rolling elements (balls); d_b for rolling element (ball) diameter; α, β, ξ for influence coefficients, defined by the wheel and shaft design; a for distance from the left support to the wheel; b for distance between the shaft supports; G for wheel weight; EI_1 and EI_2 for flexural stiffness of the shaft section; J_X, J_Y, J_Z for inertia of WM along the corresponding axes; J_E for RWM equatorial moment of inertia; l for RWM's center-of-gravity shift in relation to its suspension point; ε for angle of deviation between the axis of principal moment of inertia and the spinning axis; μ for ratio of axial and equatorial moments of inertia; D for diameter of the ball bearing outer ring ; d for diameter of the ball bearing inner ring ; z for quantity of rolling elements (balls); d_b for rolling element (ball) diameter; q for contact angle of the rolling elements.

The value of the critical angular velocity of the reaction wheel is defined by the dimensions of the design elements that belong to the Equation 12. When calculating this parameter, it is important to fulfill the following condition: the value shall be outside the range of operating angular velocities of the reaction wheel [2, 3].

An operating RWM is generally a hydraulic motor installed into the cardan suspension. That is why it is necessary to consider induced oscillations [4] that occur along the two nominal orthogonal axes of RWM. Nutation oscillations are defined relatively to the axes, refer to expressions (Eq. 13–14).

Amplitudes of induces oscillations of RWM can be great in case of resonance with denominator of trigonometric values of expressions (Eq. 13–14) turning to zero:

$$(Gl - J_B\Omega^2)(Gl - J_C\Omega^2) - (J\Omega^2)^2 = 0\,. \tag{22}$$

Then the angular velocity Ω of the wheel spin, when resonance occurs, is defined by expression (Eq. 15).

In this expression μ is the coefficient defined as relation of the RW moments of inertia along the three orthogonal axes:

$$\mu^2 = \frac{J_Z^2}{J_X J_Y}\,. \tag{23}$$

Analysis of expression (Eq. 14) shows that with $\mu<1$ the radicand will always have two positive values. Consequently, in this case, the increase of the rotor self-rotation angular velocity Ω results in double vanishing of denominator in expressions (Eq. 13–14), which is indicative of two resonant values for Ω.

If $\mu>1$, the denominator of expressions (Eq. 13–14) will go to zero only with one of the angular velocity values Ω. Thus, at the rotor run-up RWM will go though resonance only once.

In the limiting case, with $\mu=1$, the denominator of expressions (Eq. 13–14) will also vanish when $\Omega=\infty$.

Expressions (Eq. 13–14) demonstrate that the amplitudes of induced oscillations go to zero with $\Omega=0$.

The RWM wheel is usually supported by ball bearings. The parameters for RWM with ball bearing supports are mainly governed by the structural dimensions of the ball bearing elements (Eq. 16–21), as well as relation of axial and equatorial moments of inertia attributable to the RWM rotating parts (reaction wheel, electric motor rotor, reaction wheel shaft (Eq. 14–15)) [4, 5].

It is obvious that the mathematical model of RWM is quite sophisticated and cannot be based on calculation expressions. Calculation stage of RWM designing process should also include modelling and analysis with the assistance of up-to-date CAD systems.

At the present moment there is set trend, established in the world of space vehicles designing: WM weigh decrease, on the one hand, and provision of proper strength characteristics, on the other hand. That is why integrated method is the optimum way to improve the quality of frequency response characteristics of RWM.

When designing RWM the following stages of the integrated method must be observed [6]:

I. calculation using mathematical model;

II. designing and prototyping;

III. experimental run up of the prototype;

IV. analysis of the results and correcting the engineering documentation.

One of the structural schemes to implement this method applied to the design of the space vehicle attitude control system is shown in Figure 2.

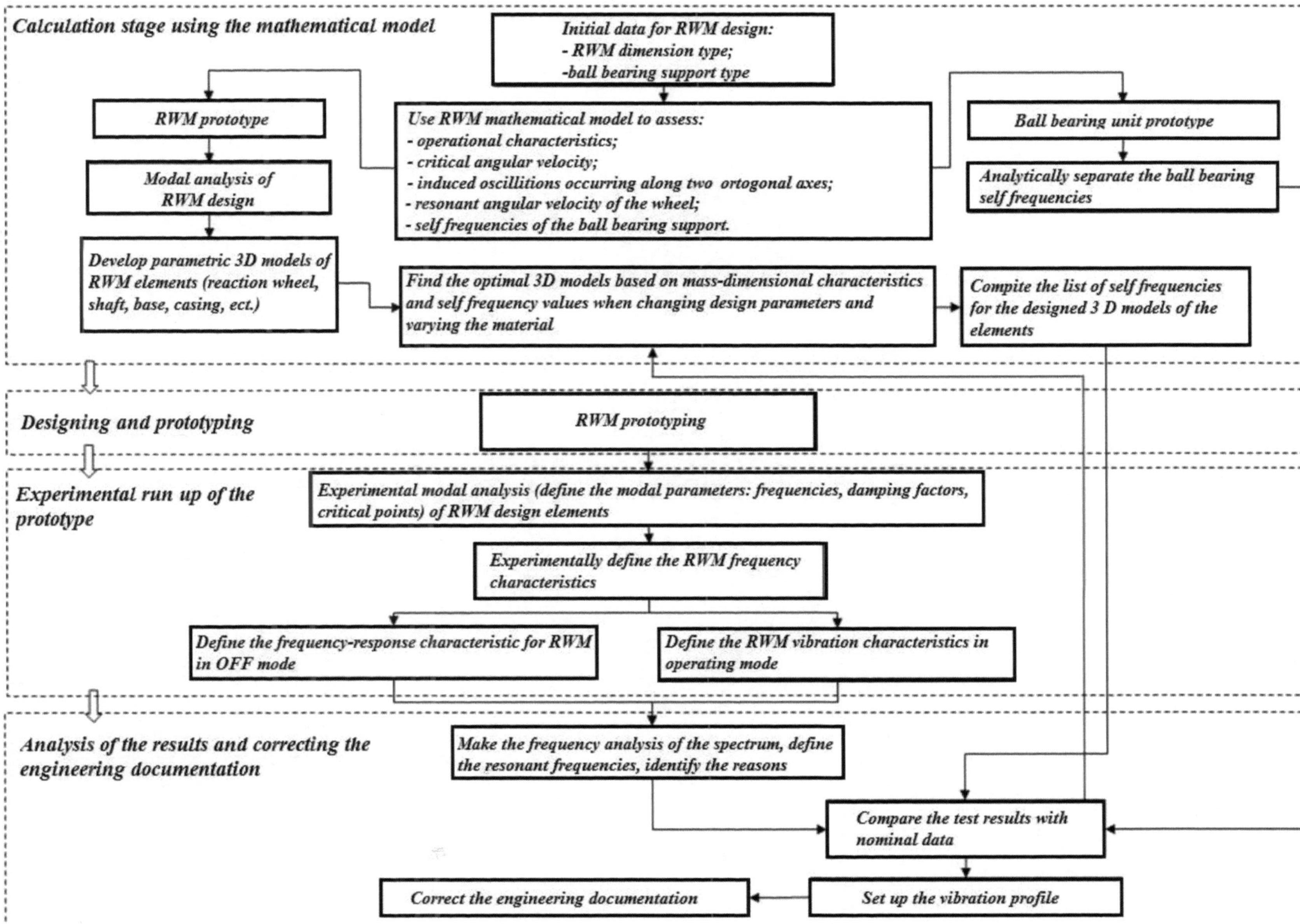

Figure 2. Structural scheme of the integrated method implementation

The proposed integrated method has a number of specific features: taking into account the experience of previous engineering works, interconnecting of the RWM calculation and engineering with RWM prototyping and test run-ups.

When seeking the RWM weigh minimization, we shall correlate its value with RWM mathematical model parameters, in order that the stiffness of the RWM in general and, consequently, the self frequencies of the system parts, the shift of the critical velocity to the reaction wheel angular velocity range, are not affected by, for instance, decreased sections of load-bearing elements, addition of supplementary weight reduction openings, etc [7].

And the most effective way to optimize the whole set of characteristics is to vary the system parameters while designing it.

Results

The results of integrated method application when designing RWM are shown in Figure 3. Here we can see parametrical 3D models of RWM main elements - reaction wheel (*a*), base (*b*) and casing (*c*) with the forms of oscillations associated with the first self-frequencies [8].

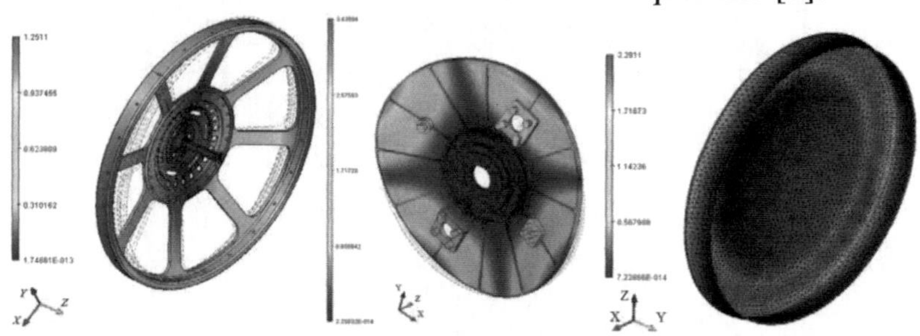

a *b* *c*

Figure 3. Forms of oscillations of RWM parts at the first self-frequency
a – reaction wheel; *b* – base; *c* – casing

The points cited above indicate that the RWM designing is quite a complicated technical task involving the need to combine dozens of operating parameters.

The frequency analysis of designed 3D models of the system elements enables to make the necessary corrections to the engineering documents at the stage of designing, thus complying with the task of satisfying the specifications.

Conclusion

The requirements imposed nowadays on the set of executive element characteristics dictate the necessity to naturally fit their mathematical models (via dedicated CAD systems) into the stage of RWM design. Literally, the whole technical level is formed at the design stage.

The closed-loop cycle of this set of theoretical and experimental researches allows formulating an integrated method aimed at improvement of the frequency response characteristics of executive devices and, as a consequence, of the quality of tasks fulfilled by the space vehicle.

The universal character of the developed method expands the number of application opportunities when designing any complex technical systems.

References

[1] G.N. Gladyshev, V.S. Dmitriev, V.I. Kopytov. Space vehicle control systems training manual, Tomsk, 2000.

[2] E.J. Heide, R. Ferreira, P. Put, P. Le. Reaction wheels at Bradford Engineering // European Space Agency. 8th International ESA Conference on Guidance, Navigation & Control Systems. 2011. C. 1-10.

[3] Yu.A. Britova, V.S. Dmitiriev, A.A. Vasiltsov, I.S. Kostarev. Influence of rigidity of power

elements of a design on size of critical speed of an executive office on the basis of the engine-flywheel operated on speed, Journal of Control. Diagnostics, 2012, vol. 11. Available at: http://www.td-j.ru/index.php/archive/216-11.

[4] X. Guan, G-T. Zheng, Q.-W. Wang. A pseudo-active vibration isolation method for reaction wheels // Yuhang Xuebao/Journal of Astronautics. 2010. T. 31. № 7. C. 1870-1876.

[5] M. Wagner, S. Airey, G. Piret, P. Le. European space agency – ESA new reaction wheel characterization test facility (RCF) // 35th Annual as guidance and control conference. 2012.

[6] D.J. Ewins. Modal Testing – theory, practice and application. ISBN 0-86380-218-4. Research Studies Press Ltd. 562 p.

[7] Leung Y.T. Dynamic substructure response. // Journal of sound and vibration. -1991, vol. 149. p.83-90.

[8] ME'scopeVES 4.0. Operating Manual. Volume I – TOTURIAL // Vibrant technology, Inc. – 2003. – 210 p. 2010. URL: http://www.vibetech.com/assets/mescope/MEscopeVESVolI.pdf.

Experimental Research of High-Energy Capabilities of Material Recognition by Dual-Energy Method for the Low-Dose Radiation

A Abashkin[1], S Osipov[2], S Chakhlov[3] and A Shteyn[4]

[1]Student, National Research Tomsk Polytechnic University, Tomsk, Russia
[2]Leading Researcher, National Research Tomsk Polytechnic University, Tomsk, Russia
[3]Head of Laboratory, National Research Tomsk Polytechnic University, Tomsk, Russia
[4]Engineer, National Research Tomsk Polytechnic University, Tomsk, Russia

E-mail: anthony.abashkin@gmail.com

Abstract. The algorithm to produce primary radiographs, its transformation by dual energy method and recognition of the object materials were enhanced based on the analysis of experimental results. The experiments were carried out at the inspection complex with high X-ray source – betatron MIB 4/9 in Tomsk Polytechnic University. For the reduced X-ray dose rate, the possibility of recognition of the object materials with thickness from 20 to 120 g/cm^2 was proved under the condition that as the dose rate is reduced by the defined number of times, the segment of the image fragment with the reliably identified material will increase by the same number of times.

1. Introduction

Recognition of materials of the object of inspection control and its fragments remains one of the most important problems to be solved by the customs and border authorities in different countries all over the world, security services, passenger and freight traffic services. The recognition of materials applied to the inspection control is traditionally understood as the comparison of materials the object is made of or fragments thereof by the effective atomic number and its associated parameters to one of the broad classes of compounds. All materials are divided into a number of classes, wherein each of the classes related to its range of variation of the effective atomic number is associated with its most common representative. The class of organic materials includes, for example, water, graphite, polyethylene, polypropylene and others, any of these materials can be selected as a typical representative of the class. A class of materials with a low effective atomic number is associated with aluminum. A typical representative of the class of metals and alloys with an average effective atomic number is steel. A class of materials with a high effective atomic number is associated with lead. At present, the recognition of materials of the inspection control of objects and their fragments are widely used for various implementations of the dual-energy method [1–7]. The inspection systems based on the betatron and linear electron accelerators are used to control large objects [8–11]. One possible way to improve the performance of inspection control is to scan vehicles with drivers. This approach leads to the formation of primary radiographic images for the object of inspection control at two levels of radiation dose: low level – to scan the cab with the driver; high level – to scan the rest of the vehicle. The low level is specified by the radiation safety standards. This implies the needs for experimental verification of the possibility of recognition materials and their fragments by high-energy dual-energy method for low levels of X-ray radiation dose.

Initial radiographic images were obtained using the inspection complex in Tomsk Polytechnic University for the energy pair 4 MeV–7.5 MeV. The inspection of test objects made of organic materials (water, plexiglas, nylon, wood), aluminum, carbon steel and lead was carried out to the mass range of 20 to 120 g/cm^2. The test object with the scanner is placed in the most problematic area – at a maximum distance from the axis of the beam of high-energy X-rays. For more accurate estimation of the parameters of calibration line, test objects are scanned at a speed of 4 cm/sec. The images were formed in the mode 1 low energy pulse and 1 high energy pulse. To confirm the basic hypothesis, at first we need to estimate the range of radiometric signal levels while reducing the absorbed dose of radiation, and then to investigate how the change affects the quality of recognition of materials of different classes – organic, aluminum, steel and lead.

2. The range of changes in the levels of radiometric signals reducing the absorbed radiation dose

Radiometric level signal is proportionally reduced with the decrease in dose rate. Deviation from this law is caused by two factors: inaccurate measurement of the absorbed dose rate and limitation of the betatron monitor by bit ADC. During the research the exposure dose of radiation beam axis P is reduced from 2.5 to 0.3 R/min. The dose rate in the scanning area is about 3.8 times less than in the center. Table 1 shows the values of radial thickness R1 and identification parameter Q. To simplify the analysis of the results presented in Table 1, the following color palette was used with the proper material recognizing fragments: blue – organic materials; green – aluminum and alloys; red – iron alloys; yellow – lead alloys. Incorrect recognition is marked in black. Averaging is performed over the entire area of the image fragment without boundary effects.

Table 1. Recognition of test materials fragments of objects while reducing the radiation dose.

Z	Material	*P*, [R/min] (center and periphery)							
		2.5 (0.665)		1 (0.266)		0.5 (0.133)		0.2–0.3 (0.053–0.08)	
		R_1	Q	R_1	Q	R_1	Q	R_1	Q
6	Polyamide (caprolon)	0.8358	0.74173	0.8304	0.74150	0.6948	0.67544	–	–
		1.5147	0.74150	1.5080	0.74171	1.3931	0.72513	–	–
		2.2135	0.73458	2.1776	0.73746	2.1611	0.73467	–	–
		3.3298	0.73276	3.2649	0.73820	3.2226	0.74645	–	–
	Tree (pine)	0.6847	0.74297	0.6775	0.74283	0.6824	0.74057	0.6749	0.76409
		1.2380	0.74122	1.2468	0.74158	1.2332	0.74395	1.2394	0.74658
		1.7785	0.73880	1.8002	0.73489	1.7643	0.74534	1.7477	0.74874
		2.2766	0.73612	2.2519	0.74585	2.2557	0.74369	2.2872	0.72867
		2.7822	0.73275	2.8065	0.73095	2.8207	0.72422	2.7406	0.73966
		3.2438	0.73483	3.3448	0.71376	3.2861	0.72531	3.1274	0.75144
13	Al	0.8433	0.76100	0.8468	0.76452	0.8095	0.76778	0.8280	0.77502
		1.5920	0.75800	1.6059	0.75613	1.5839	0.76285	1.5709	0.76999
		3.4022	0.76070	3.4525	0.77793	3.3567	0.78681	3.4578	0.77751
26	Fe	0.4861	0.77592	0.4858	0.76862	0.4882	0.76582	0.4860	0.77617
		1.1349	0.77681	1.1341	0.77426	1.1392	0.77319	1.1355	0.78386
		2.1109	0.78361	2.1142	0.78049	2.1050	0.78572	2.1058	0.78570
		2.9786	0.79258	2.9546	0.79686	2.9447	0.80301	2.9406	0.79857
		3.7536	0.80897	3.7200	0.81339	3.5561	0.8460	3.4689	0.87406
		4.3138	0.84531	4.2839	0.84658	4.1478	0.87807	3.7423	0.96352
		4.9245	0.85487	4.7050	0.89355	4.2663	0.98441	3.9855	1.03136

82	Pb	1.5175	0.79454	1.5376	0.78604	1.5259	0.79097	–	–
		2.6269	0.83795	2.5956	0.804603	2.4109	0.90924	–	–
		3.0701	0.85022	3.0456	0.85394	3.0777	0.84960	–	–
		3.5279	0.87392	3.4862	0.88427	3.5538	0.87221	–	–
		3.9120	0.89282	3.9170	0.88615	3.6936	0.87787	–	–
		4.5821	0.91246	4.4539	0.93654	4.3531	0.94519	–	–
		4.8774	0.95578	4.6305	1.00448	4.5391	1.01160	–	–

The analysis of the data from Table 1 makes it possible to conclude that the range of mass thickness fragments, for which you can confidently recognize the material of control object with the decrease in dose rate radiation, is narrow. When $P \approx 0.266$ R/min, all the analyzed materials are confidently recognized in the range of mass thicknesses from 20 to 100 g/cm^2.

3. Experimental evaluation of recognition quality for the materials by dual-energy method (complex test objects)

The quality of recognition of the following materials was assessed experimentally: organic materials – flour, plexiglass, water; light metals and inorganic materials – aluminum, cement, salt; metals with the average effective atomic number – steel, copper; heavy metals – lead. The mass thickness of fragments was close to 40 g/cm^2. A total amount of 9 scan cycles for test objects were performed. In each cycle the dose rate varied –3 R/min; 2 R/min; 1 R/min; 0.5 R/min. Test objects from cycle to cycle had different mutual arrangement of fragments of various materials in height. The lower detector unit operated at lower levels of radiation dose rate than the central blocks. To illustrate this, the final images of the dual-energy method of test objects for 1 compilation are summarized in Table 2.

Table 2. The final images of test objects for 1 compilation.

Object	P_{max} [R/min]	Image
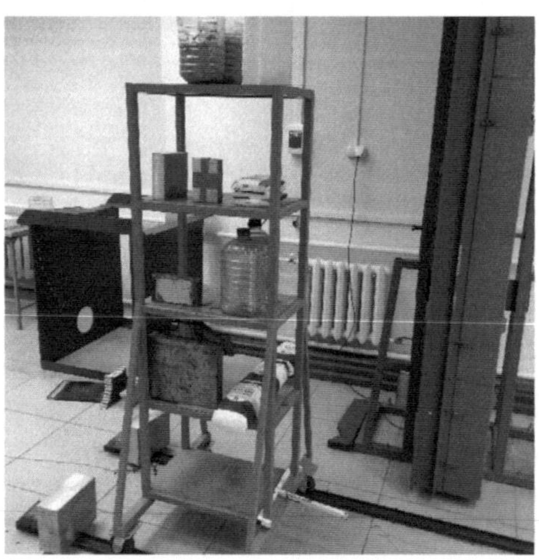	3	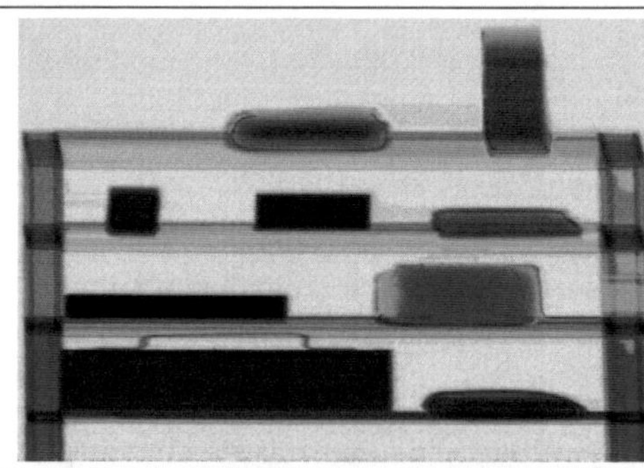
	2	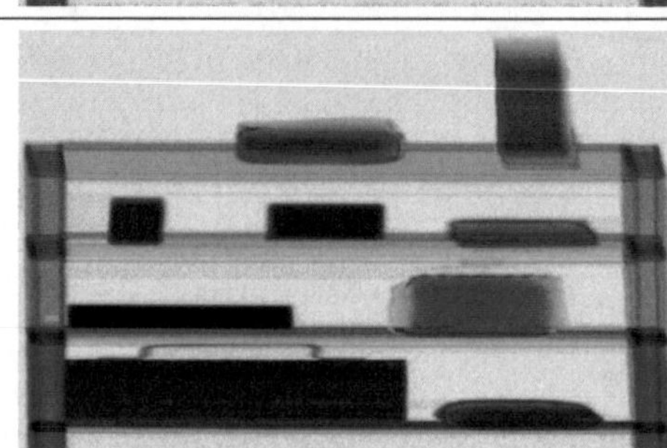

1	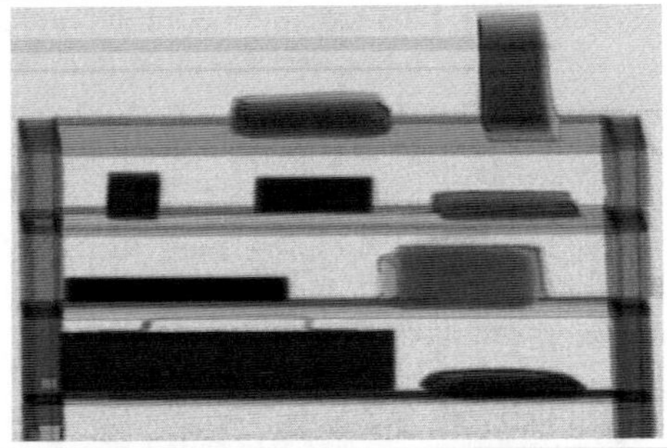
0.5	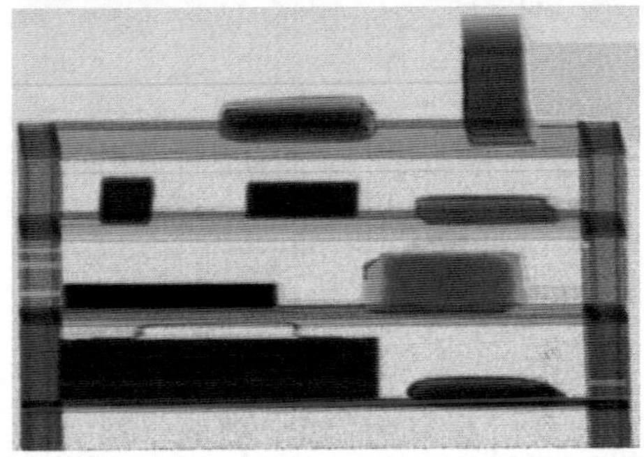

Images were obtained using a modified program Dual averaging window with 5 × 5 pixels, the minimum cluster size – 10 × 10 pixels. The images are painted according to the color scheme used by Heimann. A number of conclusions can be made based on the image analysis presented in Table 2.

1. With decreasing the dose noise levels of semi-tone pictures and color noise increase;
2. With a decrease in dose rate a significant part of the color fragments is disordered;
3. The distortions are more precise for the lower part of the image;
4. The boundary effects due to averaging and clustering appear mostly for pieces at the top of the test objects;
5. The size of the fragment, the material which is detected at the predetermined probability depends on the object's position by the vertical axis and its size in the direction of X-ray photons.

Table 3 shows the results of quantitative and qualitative detection of fragments of testing materials. The table shows the average values of effective atomic number of the material, the color to indicate numbers, and the class of materials.

Table 3. Quantitative and qualitative results of the detection of material fragments of test objects.

Material	Dose of radiation, [R/min]				Comment
	3	2	1	0.5	
Flour	10.2	10.4	10.4	10.8	Small size by vertical axis
Caprolon	4.5	4.5	4.0	4.0	
Water	6.0	6.3	4.9	6.1	
Cement	18.0	19.8	19.4	21.4	Correction recognition
Salt	16.4	16.3	16.0	19.4	
Aluminium	16.3	18.0	14.3	16.5	
Steel	24.9	25.7	24.0	26.3	
Copper	31.7	32.9	32.8	33.1	
Lead	51.9	53.3	49.5	52.2	

The results of the analysis shown in Table 3 prove the possibility to use the data on the effective atomic number of the material. Increase in the size of flour fragments in the scan direction substantially reducing the boundary effect leads to the correct recognition of the material fragments. The fact that the dose of X-radiation at the periphery is 2–3 times lower than in the middle of the beam allows making a conclusion about accurate recognition of the objects of material fragments of control with the decrease in dose levels of 0.25–0.3 to R/min. The information considered above is valid for proper cross fragment size of control objects and object fragments with thickness of about 40 g/cm^2.

4. Conclusion

The paper shows the main problems of recognition of materials of the object of inspection control by dual-energy associated with a decrease in the absorbed dose of radiation. It was proved that the radiometric range of the signal varies considerably, thereby reducing the thickness of the fragments, materials which are recognized with the given probability. This factor is the most significant for peripheral blocks of radiometric detector located at a considerable distance from the axis of the X-ray beam. The suggested invariant recognition method links the recognition of quality control and performance of testing object fragment area, the material which is recognized with the given probability. The experiments proved the possibility of a satisfactory recognition of materials of the object of inspection control at power levels of X-ray dose of about 0.25 R/min. The significant influence of boundary effects on the quality of recognition materials has been proved.

Acknowledgement

The authors acknowledge the financial support of the Ministry of Education and Science of the Russian Federation.

References

[1] Kolkoori S *et al* 2014 *Proc. of 11th European Conference on Non-Destructive Testing (Prague)*
[2] Michel S *et al* 2014 *Int. J. Ind. Ergonom.* **44** 551–560
[3] Tuszynski J *et al* 2013 *J. Transp. Secur.* **6** 339–356
[4] Mery D *et al* 2013 *Proc. of Computer Vision and Pattern Recognition Workshops (Portland)* 368–374
[5] Uroukov I *and* Speller R 2015 *Signal Process. Res.* **4** 1–11 DOI: 10.14355/spr.2015.04.001
[6] Vasendina E A *et al* 2015 *IOP Conference Series: Materials Science and Engineering* **81 (1)** 012070 DOI:10.1088/1757-899X/81/1/012070
[7] Payuk L A *et al* 2016 *Journal of Physics: Conference Series* **671 (1)** 012044 DOI:10.1088/1742-6596/671/1/012044
[8] Osipov S P *et al* 2015 *IOP Conference Series: Materials Science and Engineering* **81 (1)** 012082 1–13 DOI: 10.1088/1757-899X/81/1/012082
[9] Nedavnii O I *et al* 2002 *Rus. J. Nondestruct. Testing.* **38** 170–175
[10] Nedavnii I O *and* Osipov S P 2015 *Rus. J. Nondestruct. Testing.* **41** 260–265
[11] Osipov S P *et al* 2014 *Rus. J. Nondestruct. Testing.* **50** 491–498

IOP Conf. Series: Materials Science and Engineering **132** (2016) 012013 doi:10.1088/1757-899X/132/1/012013

Diagnosing the Bottom of the Vertical Tank, a Device Based on the Phased Acoustic Gratings

Yu Bezborodov[1], A Selsky[2], N Lysyannikova[3], Ye Kravtsova[3], A Lysyannikov[3], V Shram[3], Yu Kaiser[3] and M Merko[3]

[1]Head of Department, Siberian Federal University, Krasnoyarsk, Russia
[2]Leading Specialist, NK NOU «NUC «RTS», Krasnoyarsk, Russia
[3]Associate Professor, Siberian Federal University, Krasnoyarsk, Russia

E-mail: labsm@mail.ru

Abstract. This paper presents a decision on the creation of a specialized low-frequency flaw detector operating on acoustic phased arrays. A defectoscope will solve the problem of oil leakage through diagnostics of installation or operational defects of the bottom base metal sheets that are difficult to detect by conventional methods of testing due to a large area of the object.

1. Introduction

The most dangerous type of wear in the operation of vertical steel tanks is the formation of leaks in the leaf elements of the walls and bottoms due to corrosion under exposure to the external environment and possible contamination (precipitation) in the stored petroleum product. The adverse effects of leaks are loss of oil, increased fire risk, environmental pollution and undermining the base of the tank [1]. In this regard, the "Rules of technical operation of tank farms" provides periodic measures for tank diagnosis.

In the majority of typical methods for control of the tank bottom, attention is paid to the inspection of welded joints. However, the welded connection of the heads during operation of the tank is not exposed to high tensile loads, and therefore, it they may be tested for tightness only. On the bottom made of rolls welded by automatic and semi-automatic welding, inspection by ultrasound [2] should be performed only for mounting seams, and such welding defects as non-through the fusion, undercuts, pores and removing toxins are unlikely to pose any danger. However, much more actual control is that of the base metal sheets. For example, according to statistics of weld observations, not more than 10% of the total detectable leaks are in the bottoms of tanks [2, 3].

Distinctive features of the exploitation of oil reservoirs in terms of the mechanisms of wear of the walls and bottoms are loading and the predominance of chemical effects of the environment on the activity of the stored product.

In view of the accessibility of leaks in the vertical walls of the test tank and in the course of their operation for visual inspection, their detection does not pose practical difficulties, and chemical effects of the environment on the metal walls are predominantly one-sided (outside) and less active than those on the bottom. Visual detection of leaks in the bottom with full tanks is impossible.

Removing the static pressure significantly alters the current state of the defects in the bottoms. Through or penetration defect occurring potentially under pressure from the liquid column gets crack opening sufficient for penetration of the product, but when unloading the tank under the action of elasticity of the metal it can be drawn together to the extent that it is able to detect defects visually. Operational defects arising from the soil are available for visual identification regardless of their size.

2. Results and Discussion

Typical methods of the control of the bottoms of tanks listed in the instructions, provide for selective testing of visually suspicious places with ultrasonics (weld and thickness measurements), and vacuum-liquid method (through defects). The vacuum-liquid method implies that a suspicious part moistened with soapy water is placed in a transparent vacuum chamber (hood Plexiglas), with the air evacuated. Under atmospheric pressure, the air rushes from the bottom into the camera through the defect, and therefore, the defect forms bubbles [4, 5].

However, such methods do not solve a number of problems, namely: they do not reveal defects in the form of ulcers and non-through cracks that develop from the soil, and also through the "collapsing" cracks that cannot be distinguished visually because of their weak erections when unloaded RVS. To solve this problem a special technique of continuous ultrasonic flaw detection was developed using normal acoustic waves in the leaves (Lamb waves). This enabled, for example, full control of the bottom vertical tank for storing diesel fuel with a diameter of 18 m (an area of 2550 m^2) with the cost of finding defects of 8 man-hours. However, the drawback of this technique is that it is focused on the parameters of the model universal ultrasonic flaw detectors, manufactured by domestic (UD2-12, UD2-17) and foreign (USK-7S) industry at the time of its development, which do not always most correctly satisfy this task.

This primarily refers to a set of ultrasonic frequencies stipulated by the standards for such equipment. The most universal flaw detectors use frequency below 1 MHz, which limits the "range" of the acoustic signal as the frequency increases, and its tendency to fade increases. The developed method works for the "range" of 1 m. If we start not from the universal instrument and the conditions of the problem, the optimal frequency should be considerably lower than that realized in the method versions (1–2 MHz). Moreover, in recent years, in the global market of flaw detection equipment, computerized models with low ultrasonic frequencies such as UD3-103 "Peleng", UDS 2-52 "Probe" and USM-22L appeared.

However, the selection of hardware is only a part of the solution of the technological tasks of ultrasonic diagnostics of individual facilities. Each task of this type requires optimal selection, and at the best case, it is special design and manufacture of the piezoelectric transducer (sensor) for the most efficient transfer of oscillatory pulses of electric energy into ultrasonic probing signals of elastic oscillations of a selected type in a specific metal object and return of the acoustoelectric conversion echo from all possible defects.

Currently, many developers of ultrasonic flaw detectors turned to the use of the so-called PAAG-converters ("PAAG" is phased array acoustic grating). They are convenient to use since the insertion angle can be changed automatically, that is, it scans a swinging beam. The general scheme and principle of operation of phased arrays is shown in figure 1.

A phased array radar is piezoplasty frequent cut into narrow strips (width is approximately equal to the length of the longitudinal wave in the material, which is pasted on the lattice). Each lane corresponds to an individual channel for radiation-reception of the inter-channel phase shift, the value of which is adjusted automatically. The principle of operation of the PAAG is based on the common mode interference of waves (figure 1). If the phase shift between the fronts of both bands is zero, the same edges 1a and 1b and the following one, spherically spreading, will be in phase overlap between the strips at the point sliding straight down. At this point, the signal is always twice as weak as in the other points of the single-phase fronts. Therefore, its trajectory along the total acoustic axis is shown for two bands. If the phase shift is not zero, the intersection points are the same, and the fronts are moving. The greater the phase shift, the greater the angle of the beam. The increase in the number of lanes expands the wave flow.

The phase shift between the radiation channel and the reception one is created in the instrument using electronic circuits made of simple elements called delay lines (two resistors and a variable capacitor, which determines the shift value).

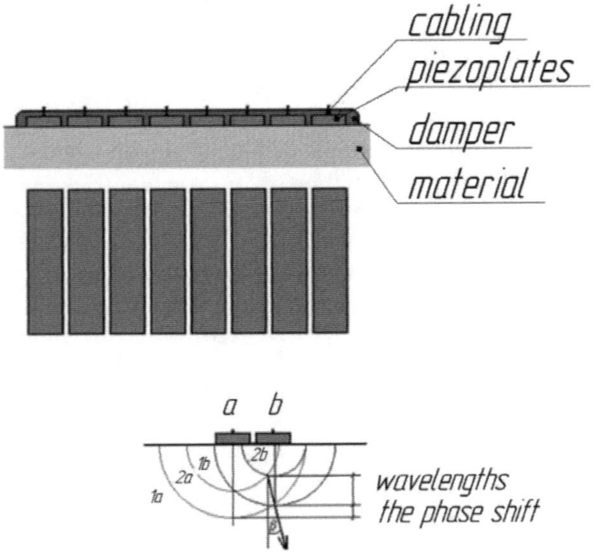

Figure 1. The general principle of operation of phased-array acoustic grating operating in a vertical plane.

In the vertically operating PAAG, a converter in the material, only longitudinal waves are created under the grate (figure 2). Since the source of these waves is the material surface, the transformation into a transverse wave occurs on it. In case of a plane-parallel object, it will occur during the first intermediate reflection from the bottom surface. Under equal conditions, longitudinal waves are less sensitive to small defects than transverse ones, they also they have the property of partial passing through closed planar defects and are weakly reflected from the right angle between the crack and the surface of the object. Therefore, this option is suitable to test coarse-grained materials only (e.g., austenitic steel) and only a direct beam without intermediate reflections. For the generation of flow shear waves in the design of the PAAG, a converter should be a plastic prism (figure 2). In this case, the angular limits of the beam oscillation in the prism are first and second critical angles that are set when configuring the instrument. Receiving echo signals of the grating also use the phase shift: the wave alternately rolled in all piezopolaron and the lattice, and the same delay lines allow you to fold the amplitude of the echo and to take this sum as the total amplitude of the reflection.

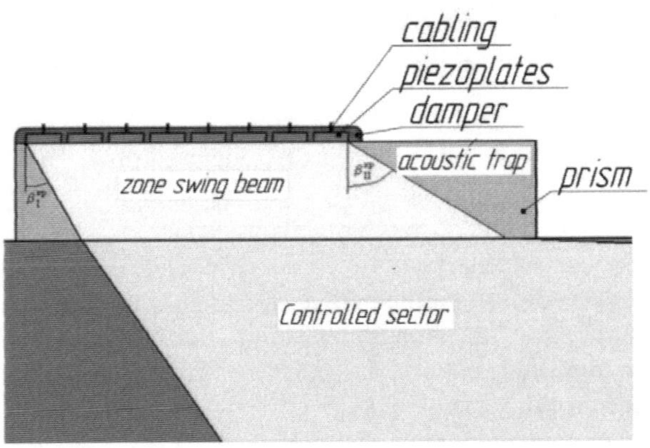

Figure 2. The scheme vertically current PAAG - converter.

In PAAG-detectors these processes occur at each value of the angle input and automatically modify with a very small step. The frequency change of the angle set by hardware is not greater than the frequency of the signal packages so that the pattern of the reflections generated on the screen of the device after accumulating and processing all the current data guaranteed the time to emerge at each angular value. Thus, the scan type "B" (profile) generated on the screen is a sector in which the sensitivity is the same in all directions, and the defects are represented as colored spots, approximate images. In these images, the color varies depending on the amplitude of the echo, and with the help of auxiliary devices (mouse, cursors, etc.) we can determine the numerical values of all required defect parameters (amplitude, coordinates, conditional height).

In the developed detector, AC signal phasing (automatic change of the azimuth) should be carried out not in vertical and horizontal planes, so as to excite lamb waves, the vertical angle of the ultrasound needs to be fixed, and automatic change should be subjected to the azimuth direction of the signal in the controlled plane of the sheet. However, note that technically this phase can be performed only within the angle from 0 to 90°. Therefore, the design of the PAAG, where a converter should consist of 4 gratings arranged in a square plane (figure 3), and built-in flaw in the program, respectively, to represent the image of a leaf with marks of detectable defects on the screen.

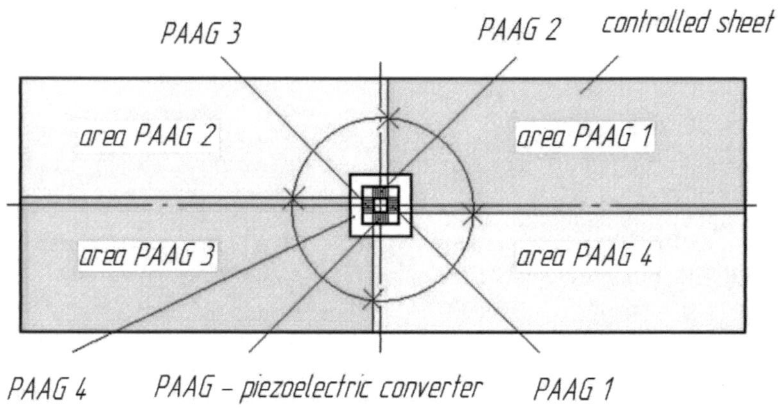

Figure 3. General scheme of an inspection of steel sheet PAAG - system.

Phase representation of signals in electronic circuits is performed using delay lines, a schematic diagram of which is shown in figure 4. Automatic change of the magnitude of the phase shift (delay signal) is achieved by simultaneous change of the potential resistance $R\phi$ or the capacitance of the S_F. The number of sections of delay line 1 is less than the number of phased array elements, as the first element of the PAAG receives a signal received at the input of the delay line, and the other receives it from the outputs of the respective sections.

The algorithm of automatic scanning that needs to be provided with specialized software PAAG flaw detector is shown in figure 5.

Software developers of the specialized PAAG -flaw detector must meet the following conditions:

- an image on the screen should be in the form of a rectangle that simulates the contours of the tested sheet;
- a label position sensor should be fixed in the center of the scan;
- a defect item should appear as a dark spot on the respective areas at a distance from the center corresponding in scale to the coordinates of the defect;
- when you hover the mouse over the image, the defect next to it should appear in the numerical parameters of reflection: the amplitude echo in decibels and the coordinates in degrees and millimeters relative to the installation point of the sensor.

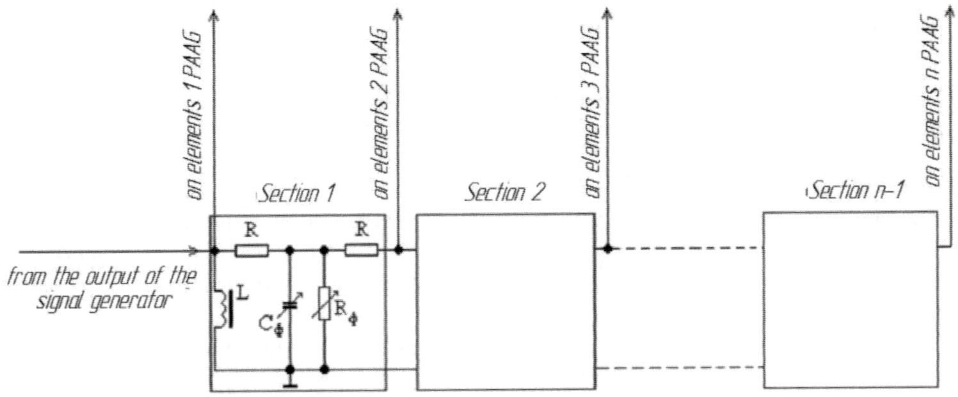

Figure 4. Schematic diagram of the delay line signal.

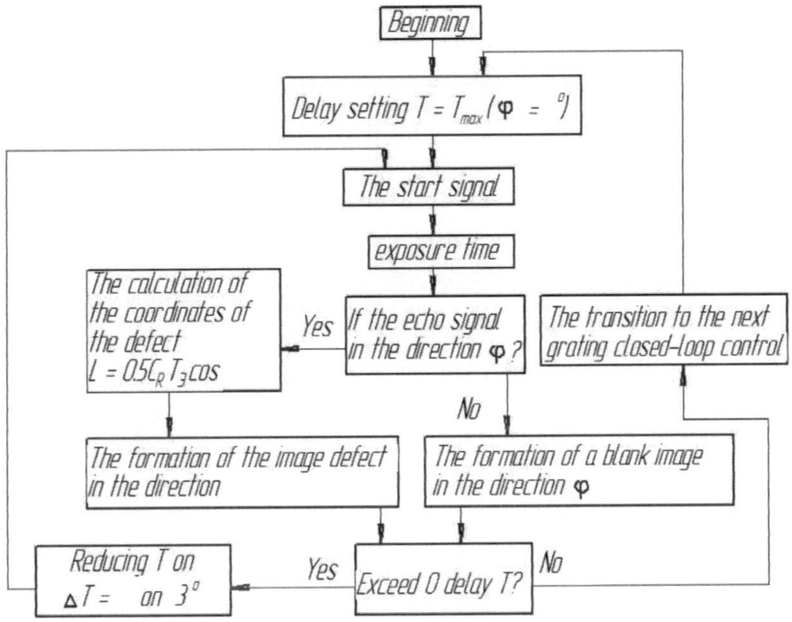

Figure 5. Algorithm auto scan.

An example of the intended image is shown in figure 6.

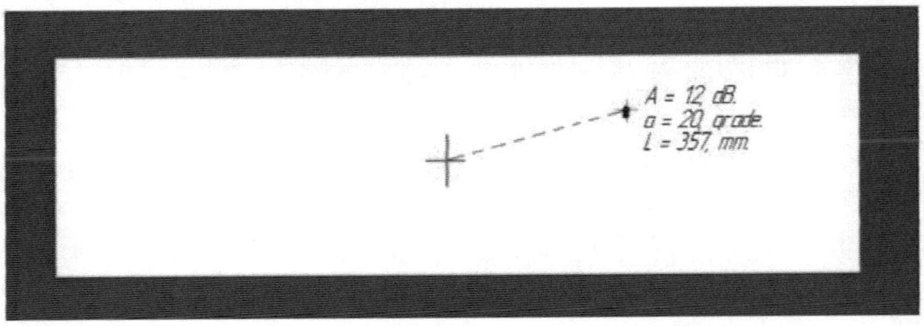

Figure 6. Example of the screen image.

IOP Conf. Series: Materials Science and Engineering **132** (2016) 012013 doi:10.1088/1757-899X/132/1/012013

3. Conclusion

The developed specialized low-frequency flaw detector operating on PAAG will solve the problem of oil product leakage by providing early diagnosis during manufacturing and installation or operational defects of the bottom base metal sheets which are difficult to identify by conventional testing methods due to a large area of the object.

References

[1] Commandant instruction M5090.9 1995 *Storage tank management manual*
[2] ASTM E164-13 2013 *Standard Practice for Contact Ultrasonic Testing of Weldments*
[3] Simpson J Blitzg, Blitz J, Blitz J and Simpson G 1996 *Ultrasonic Methods of Non-destructive Testing* 1st ed. (London: Chapman & Hall)
[4] Cartz L 1995 *Nondestructive Testing* (Materials Park, OH: ASM International)
[5] GOST (State standart) 18353-79 1987 *Nondestructive check. Classification of types and methods* (Moscow: Publishing house of standards)

IOP Conf. Series: Materials Science and Engineering **132** (2016) 012014 doi:10.1088/1757-899X/132/1/012014

Means and Method for Measurement of Drilling Fluid Properties

A Lysyannikov[1], P Kondrashov[1] and P Pavlova[2]

[1]Associate Professor, Siberian Federal University, Krasnoyarsk, Russia
[2]Assistant, Siberian Federal University, Krasnoyarsk, Russia

E-mail: lysyannikov.alek@mail.ru

Abstract. The paper addresses the problem on creation of a new design of the device for determining rheological parameters of drilling fluids and the basic requirements which it must meet. The key quantitative parameters that define the developed device are provided. The algorithm of determining the coefficient of the yield point from the rheological Shvedov-Bingam model at a relative speed of rotation of glasses from the investigated drilling fluid of 300 and 600 rpm is presented.

1. Introduction

The main rheological parameters of drilling fluids are plastic viscosity, static and the yield point.

The change in the plastic viscosity upward the flow of the drilling fluid affects the ability to remove cuttings and hydraulic losses, which in turn determines the actual hydrostatic pressure in the well.

The yield point shift affects the fluidity of the solution, as the minimum level of tension should be ensured before solution deformation.

The change in the static shear stress affects the ability of the gel formed in the drilling fluid to keep the drill cuttings particles in suspension (after the cessation of the circulation) that prevents their deposition on the bottom hole and "jamming" as a possible variant and loss of drilling tools.

If the value of the static shear stress grows, the strength of the structure of the washing liquid increases resulting in difficult starting of the pumps, creates considerable pressure on the walls of the well, which in certain rocks can lead to hydraulic fracturing when restoring circulation and worsens the conditions of cleaning of drill cuttings particles and degassing of the cleaning agent.

Optimally balanced formula will prevent the possibility of oil and gas flowing from the wellbore resulting in release of the wash solution from the well and creating an emergency situation.

Continuous monitoring of drilling fluid properties allows setting an optimum mode of drilling and correcting it to reflect changes in the geological section reducing the consumption of chemicals in the drilling fluid (the cost of 1 m^3 of drilling fluid will decline and the long term savings will follow in the progressive mode).

On the market of the devices to control the rheological parameters of drilling fluids in Russia the situation is currently as follows:

- In early 1990s, on the Russian market, there were instruments monitoring the rheological parameters of drilling fluid, and foreign products found their niche. Therefore, the Russian manufacturers of similar products had lower sales volumes and suspended their modernization, and the production almost stopped;

- in 2014, the cost of foreign devices caused by rise in the dollar significantly increased;
- import to Russia of industrial equipment from the countries of the EU and the USA was banned and unless the sanctions imposed against Russia are be extended, Russian companies will be forced to resume work on product improvement, which would be virtually impossible to undertake in the short term.

2. Results and Discussion

Therefore today to measure, e.g. the static shear stress, yield point and plastic viscosity of drilling fluid it is necessary to use one foreign apparatus or two Russian apparatus such as: SNS and rotational viscometer VSN-3 (figure 1).

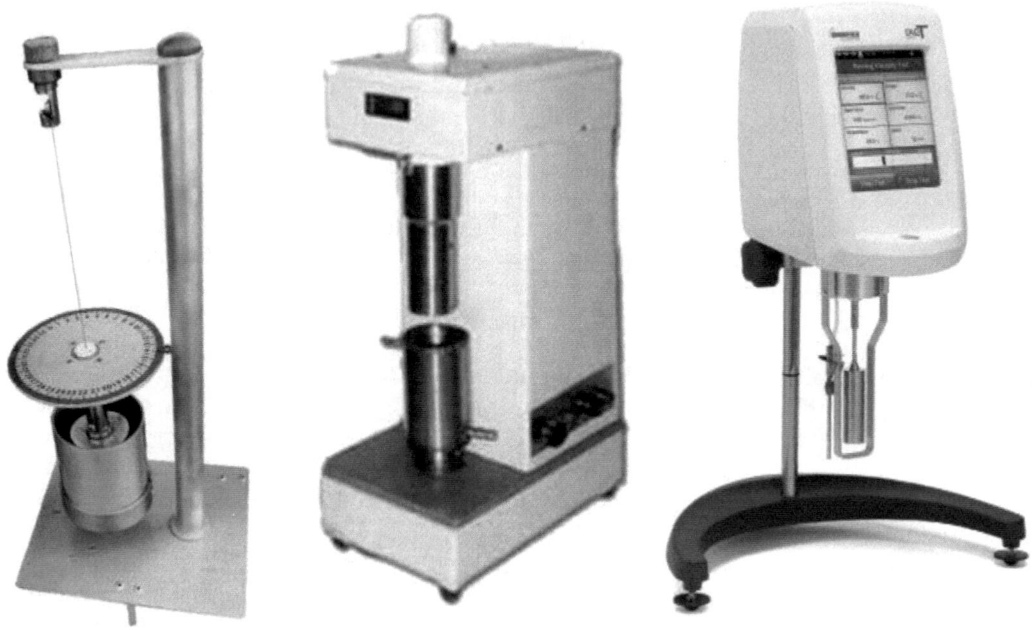

a b c

Figure 1. Apparatus for measuring the static shear stress, yield point and plastic viscosity of the drilling fluid: SNS-3 (a); VSN-3 (b); Brookfield rotational viscometer (c).

If anti-Russian sanctions are extended, the Russian companies will start production improvement. Define the key advantages and disadvantages of the existing foreign devices.

Advantages:

- high accuracy of measurements;
- the process is fully automated;
- measure the plastic viscosity (centipoise or MPa·s);
- measures the static shear stress (Pa);
- measures temperature (°C or °F);
- measure shear rate/shear stress;
- possible to set the speed of rotation of the spindle (up to 8 speeds);
- provided data visualization and display comparison;

- provided with automatic recording of results;
- automatic calibration of the device;
- ensured compliance with international standards API.

Disadvantages:

- high cost of 3 000 $;
- the complexity of the design; • in case of breakage, the inability to self-repair.
- Define key advantages and disadvantages of existing devices of the Russian production.
- Advantages:
- simplicity of design;
- easy operation;
- ability to self-calibrate;

Disadvantages:

- SNS totally mechanical device requiring considerable attention during the measurement of drilling fluid properties;
- as a series of the SNS and VSN – 3 allows to determine only one parameter of the drilling fluid;
- a large mass of VSN – 3 (22 kg).

Based on the analysis of the existing designs of devices a conclusion was made. Due to the fact that the mass of the BCH-3 is several times greater than that of the instrument for determining the static shear stress, the best technical solutions used in devices: SNS 2, SNS-2 SNS-3 and foreign devices are recommended for the development of the prototype as a base for improvement. The principle of operation for both Russian and foreign instruments is to measure the torque in the circular shear flow of the investigated solution with constant speed in a thin annular layer (in the gap between coaxially placed cylinders). The device records the moment of the resistance to the rotation of the inner cylinder of the measuring device with the test drilling fluid at different gradients of the shear rate and the calculation of the dynamic tension of the shift and plastic viscosity.

To assist Russian manufacturers, we need to develop a device with the following quality parameters:

- the ability to provide simultaneous measurement of at least three parameters of quality drilling fluids;
- high precision measurements of rheological parameters of drilling fluids;
- low cost in comparison with foreign analogues;
- a simple design that enables in case of failure of self repair;
- a fully automated device with the ability to connect to a PC that earlier in Russia was not possible;
- mass at the level of the imported samples [1].

Quantitative parameters:

- measurement of static shear stress, from 0 to 20 Pa.
- measurement of the maximum yield point in the range from 0 to 100 Pa.
- measurement of plastic viscosity in the range from 0.5 to 10·10-6 Pa·s
- stepless regulation of speed of rotation from 0.2 to 600 rpm;
- temperature measurement from 0 to +35 °C;
- temperature measurement accuracy ±0.1 °C;
- the limits of permissible reduced error of the viscometer is ± 1,5 %.

In spite of the intense use of stepping motors in high precision applications in robotics and industrial automation, they have been scarcely used to drive rotational viscometers. Instead, to drive these instruments, researchers and manufacturers have been employing dc motors (either conventional or Brushless types) or ac motors (either induction or synchronous types) [2-5]. The main advantage of stepping motors is the - precision (if applying potentials on windings stepping motors will rotate strictly at a certain angle), the cost of stepper drives, on average 1.5-2 times cheaper servos.

The new device required development of an algorithm for calculation of the parameters of the drilling fluid. Determining the limit shear stress and plastic viscosity is based on rheological models. All models have at least two parameters in one equation. Thus, the measurement task consists in solving a single equation with two unknowns. For example, the most common Shvedov-Bingham model has the form [6]:

$$\tau = \tau_0 + \mu_p \frac{dU}{dy}; \tag{1}$$

where τ_0 – yield point; μ_p – plastic viscosity; $\frac{dU}{dy}$ – the derivative of the velocity in the direction perpendicular to the direction of fluid motion.

The apparatus must find a yield point – τ_0 and plastic viscosity – μ_p. To determine two unknowns, you need two equations. In that way measurements are performed at a relative speed of rotation of the cylinders, 300 and 600 revolutions per minute. After test operation, the coefficient of the equation system obtained [7] are:

$$\begin{cases} \tau_{300} = \tau_0 + \mu_p \left.\frac{dU}{dy}\right|_{300} \\ \tau_{600} = \tau_0 + \mu_p \left.\frac{dU}{dy}\right|_{600} \end{cases}; \tag{2}$$

where τ_{300} и $\left.\frac{dU}{dy}\right|_{300}$, τ_{600} и $\left.\frac{dU}{dy}\right|_{600}$ is the yield point on the outer surface of the inner cylinder and the derivative of the velocity in the radial direction when the inner cylinder rotates at a speed of 300 and 600 rpm, respectively.

The yield points are associated with the angles of the ratios [3]:

$$\tau_{300} = \frac{\phi_{300} \cdot d^4 \cdot G}{64 \cdot R^2 \cdot h \cdot L}; \tag{3}$$

$$\tau_{600} = \frac{\phi_{300} \cdot d^4 \cdot G}{64 \cdot R^2 \cdot h \cdot L}; \tag{4}$$

where φ_{300} and φ_{600} – angles twisting steel filaments with a diameter d when the rotation speeds of the external cylinder 300 and 600 rpm, respectively; G – the shear modulus (G=79.3·10^9 Pa); R – the radius of the outer surface of the inner cylinder; h – the height of the inner cylinder; L – the length of the thread which is suspended inner cylinder.

The presented method is applicable to a linear dependence between the tangential stress on the outer surface of the inner cylinder and the derivatives of the velocity in the radial direction. To identify the parameters of other rheological models shown in figure 2, modern devices have more than two speeds that allow obtaining more data.

In order to precisely determine the coefficients of the rheological models, it is assumed that the next-generation device must have at least 8 speeds.

3. Conclusion

The main technical requirements to be met by the developed device are determined. The proposed algorithm based on the rheological model Shvedov-Bingama allows determining the yield point stress at speeds of 300 and 600 rpm.

The development of the new design of the instrument to control the rheological parameters will prevent the risk of possible deficit in this equipment in the Russian Federation. The instrument will provide determination of the three parameters of the drilling fluid (yield point and static shear stress, and plastic viscosity) resulting in no need to purchase additional hardware.

In further work, it is planned to develop the algorithm for determining the coefficients of dynamic shear stress and plastic viscosity based on different rheological models of the drilling fluid.

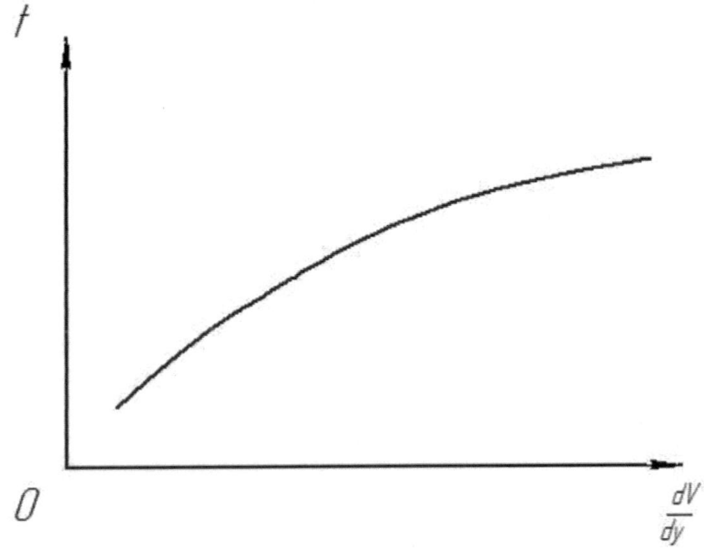

Figure 2. The dependence of the yield point.

Acknowledgements

These studies are performed in the framework of garnet CLEVER 2015-2017 year «Development of a prototype of the device to determine the plastic viscosity, static and limiting shear stress of drilling fluid».

References

[1] Lysyannikov A V *and* Kondrashov P M 2015 *Proc. of Conf. Dedicated to the 70th Anniversary of the Great Victory "Prospekt Svobodnyj – 2015" (Krasnoyarsk)* URL http://conf.sfu-kras.ru/sites/mn2015/index.html
[2] Tonmukayakul N *and* Nguyen Q D 2002 *Fuel* **81 (4)** 397–404
[3] Van Den Brule B H A A *and* Kadijk S E 1992 *J. Non-Newton. Fluid Mech.* **43 (1)** 127–139
[4] Stein D J *and* Spera F J 1998 *Rev. Sci. Instrum.* **69 (9)** 3398–3402
[5] De Lucena S E *and* Kaiser W. 2008 *IEEE Transactionson Instrumentation and Measurement* **57 (7)** 1338 DOI: 10.1109/TIM.2008.917170
[6] Tetelmin V V *and* Yazev V A 2009 *Reologiya Nefti* (Moscow: Granica)
[7] Rabia H 1985 *Oilwell Drilling Engineering: Principles and Practice* (Graham & Trotman)

Remote Diagnosis of Dug-in Areas and Bottom Pipe by Main Acoustic Emission Method Using a Self-organizing Wireless Network

Ye Kravtsova[1], V Shram[1], N Lysyannikova[1], Yu Bezborodov[2], A Selsky[3] and A Lysyannikov[1]

[1]Associate Professor, Siberian Federal University, Krasnoyarsk, Russia
[2]Head of Department, Siberian Federal University, Krasnoyarsk, Russia
[3]Leading Specialist, NK NOU «NUC «RTS», Krasnoyarsk, Russia

E-mail: rina_986@mail.ru

Abstract. This paper discusses methods of nondestructive testing, the main method is method of acoustic flue gas emission. It was found that the use of this method in the diagnosis of bottom pipe and dug-in areas allows to reduce time, does not require surface dressing to a certain value, provides almost instant information about the defect at a great distance to the nearest gas-pumping station and is not inferior to the reliability of the control other existing methods.

1. Introduction

The emergence of modern large-scale facilities - nuclear power stations, LNG terminals with, offshore drilling rigs, large chemical plants, large airliners - led along with economic benefits to large negative consequences in the event of their failure. Humanity can not refuse such facilities, but it can prevent the disaster or reduce their impact through the effective use of methods and means of nondestructive testing and technical diagnostics.

Currently, in accordance with current regulations and guidance documents [1, 2] technical diagnostics of gas mains is carried out mainly by the following methods: visual and measurement control (VIC); Magnetic (MTD) or capillary (CD) inspection; ultrasound (ultrasound) or radiation ; acoustic emission (AE); integrated control with the help of in-flaw-shells; thermal (infrared) control (TC).

There're several disadvantages in each of these methods, almost all of them (except for in-pipe and heat methods) require the immediate presence of employee on the workplace, the release of the pipeline from the insulation stripping metal, characterized by low productivity.

Control of state of the metal of dug-in areas and bottom pipe of pipeline sections are greatly complicated by the inaccessibility of objects. Only in-line flaw-shells are currently used for this purpose, which have more weight (up to 2.5 tonnes), high cost (≈ 4 million. rub.), on the long runs - low productivity (movement in collecting information no faster than0 5 m/s, seizure and analysis of the information only after the passage of the projectile and seizure).

In this paper we propose a method of using acoustic emission method, based on the reception of metal AE [3–6] signals generated by defects, their conversion into electrical signals and transmitted to the wireless receiving and recording module using self-organizing radio network. The sensors provide an autonomous low-voltage power supply, and are mounted on the inner surface of the pipeline during its assembly before going into the pond (the trench), and serve both a power signal transducers and their relays.

The main sources of emissions in metals are the motion of dislocations, plastic deformation or accompanying the emergence and growth of cracks in the structure under the influence of mechanical stresses.

Also, the source of acoustic emission (continuous) is through defects in the walls of pipes that store or transport liquid or gas under pressure, due to the friction of the flowing medium of the defect.

At the same time the surrounding metal are emitted by different types of waves: longitudinal, transverse vertically polarized (TV-waves), transversal horizontally polarized (TH-sloping wave and running parallel to the surfaces of SH-waves) Rayleigh waves, Lamb waves are normal [7].

The symbols of transverse waves: T - Transverse; V- Vertical; H - Horizontal; S - Surface.

Contact of metal with non-gaseous molecular medium (liquid, dirt and sediment, thick dense coating) causes the outflow of the acoustic energy of the metal in the environment for all of these types of waves other than SH-waves. Consequently, this type of wave has the highest "range", and that's why it would be the most profitable for the purpose of diagnosing of extended objects by the AE method. However, such wave is practically impossible to initiate or take from metal surface with a piezoelectric method.

In ferromagnetic materials it allows to realize electromagnetic acoustic method. But this method, firstly, characterized by low reduction factor of energy (not more than 5%), and secondly? requires the development of special sensors with complicated construction. Therefore, we accept the piezoelectric conversion, based on the fact that the transported medium is satisfactorily clean gas, and the thickness of the insulation coating is significantly less than the length of the sound wave AE signals and therefore do not contribute to their absorption.

Remote search for defects in bottom pipes and dug in areas of pipeline shown in figure 1. The AE signal is received from the closest sensor on the pipeline wall (receiver-transmitter module), this sensor convert it into a radio signal, which is successively relayed to other sensors and this way it delivered to the receiving antenna of registering module fixed on each gas compressor station (GCS). This provides a significant advantage, since the signal is delivered over a distance of tens and even hundreds of kilometers to the nearest GCS in a matter of seconds. Like cell phones, each sensor has its own radio frequency carrier on which is determined by the number and position of the sensor, which adopted the AE signal and sent it to the airwaves. However, even if some of the intermediate sensors are out of control, the signal will still be transmitted: based on the marginal reception range of AE signals from the metal detectors are spaced along the pipeline 20 m, and the radius of the radiation radio each of them 100 m, and therefore this signal will "hear" and transmit at least 10 adjacent sensors. This requires the introduction of broadband in the wiring diagram of the sensor unit of the relay and a second (receiving) antenna, but unlike the cell phone such additive is not particularly difficult, because unlike phones it do not require coding information transmitted in order to protect its confidentiality.

This system has a fundamental difference to the above-ground, bottom and dug-in areas of gas pipelines.

On the above-ground parts of transceivers modules are installed on the outer surface of the pipeline, the transmission of radio signals over the air goes into the surrounding atmosphere, the power - with autonomous battery charging from solar panels.

On the bottom and dug-in areas of the pipeline modules are installed on the inner surface of the pipe during installation, radio transmission is on the transported medium (ie, the pipeline itself plays the role of wave guide), power supply - cable (from the feeder, paved installation inside the pipeline). Structurally, these sensors are fundamentally different from the "above ground", as should not create resistance to movement of gas and at the same time to withstand the pressure of the environment and impact of technological "wind".

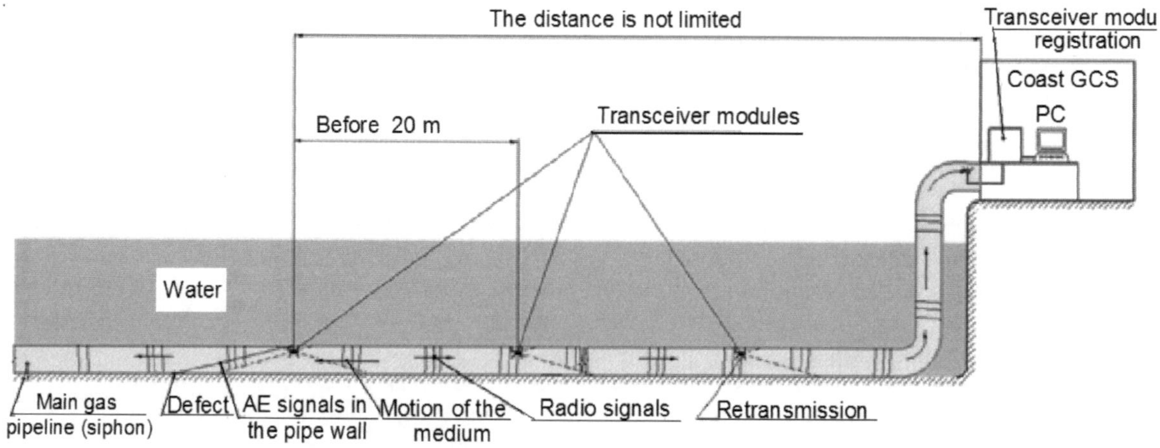

Figure 1. The general scheme of remote detection of defects in dug-in areas gas main AE wirelessly using a self-organizing radio network.

In the gas-pumping station signals are processed by the receiving-recording module and subjected to automatic analysis on computer with a special program. In addition to the basic parameters (amplitude and frequency of the acoustic emission) the following factors must be considered:

- a radio frequency on which the signal is received (the identification numbers for the receiving and transmitting unit, i.e. defective portion of the pipeline);
- a constant gain AE sensors at receiving signals from the metal and the relay (for adjusting the amplitude of the acoustic emission).

2. The principle of the reception, conversion and transmission of AE signals on the air

The generator of radio frequency carrier at its output produces uniform frequency signals for the individual sensor. These signals are sent to the first input of the comparator. If the second comparator input signal is absent then its output voltage is zero. If AE signal gets on the piezoceramic plate of the sensor, it converts into electricity and enhanced pre-amplifier. Then comparator overlays vibrations received at its inputs. As a result of lower audio frequency modulates a higher radio frequency, and the output of the comparator formed by a radio frequency signal, the contours of which are defined frequency and amplitude of the audio signal. This signal is amplified and intermediate broadband power amplifiers and a transmit antenna gets into the air (figure 2).

3. Retransmission of signals from other modules

A radio signal received from a neighboring module by the receiving antenna connected between the intermediate and output amplifier stages, amplifies the output amplifier and also via the transmitting antenna gets into the air.

4. Signal processing by receiving and registering module

The signal received by antenna amplified and gets to the input of the modulator and the first digital frequency. Frequency defines the value of the radio frequency carrier signal and delivers that information to a computer for identification numbers of the transmitting-receiving module, which received the signal.

The modulator builds envelope of a radio signal that is reshapes the AE signal. This signal gets into an analog-digital converter and a second digital amplitude frequency. ADC and a second frequency supplied to the computer data on the parameters of AE signal. As a result, computer generates information block comprising:

- the frequency of the individual modules, which adopted the AE signal;

- signal amplitude AE;
- the frequency of the AE signal.

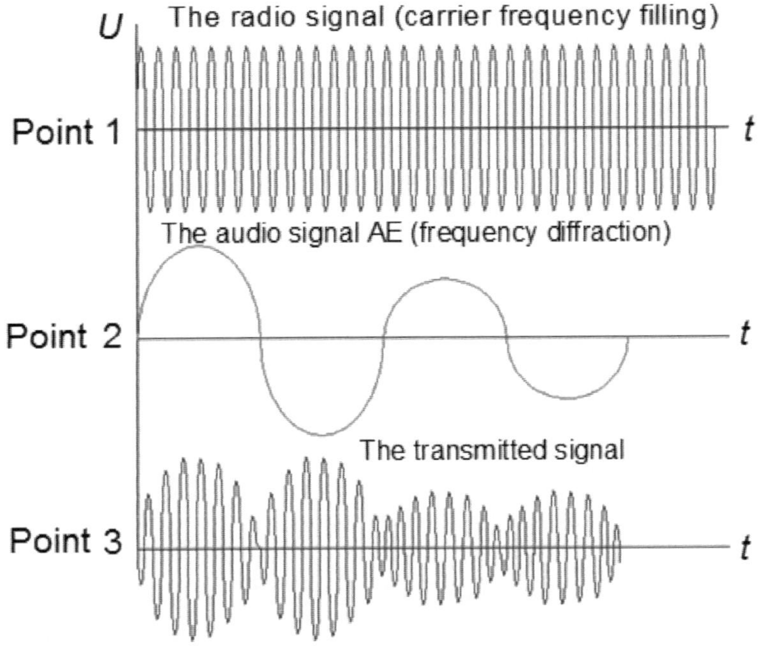

Figure 2. The principle of the reception, conversion and transmission of broadcast signal AE.

The amplitude of the signal shows the activity of the AE source (by comparison with the amplitude of the test signal, simulating a defect), and the frequency and intensity following the AE signals - about the nature of power: a continuous signal is low (less than 20 kHz) frequency - Through-flow; individual clicks on a frequency of 200 ÷ 500 kHz - structural transformation of the metal, accompanying the development of fatigue defect.

The location of AE source is determined by the number and position of receiving and transmitting unit, which accounts for the extreme value of the amplitude of the AE signals. The number and position of the unit is also automatically determined by the value of the radio frequency carrier, for which the computer must be clearly card is loaded pipeline, as well as a table of individual values of RF modules.

5. Transceiver modules and their installation

Purpose of the transceiver module is receiving AE signals from the metal of the pipe, converting them into VHF signals and sending into the air to transfer to the shore via GPS hoc network.

Requirements:

- Each of the modules should have its own radio frequency carrier indicated in the passport, as well as an individual serial number. Approximate position (geographic coordinates), number and frequency of each module during installation and downhill area in the waters have to be applied to the scheme of the pipeline;
- Distance of the effective action of typical piezoelectric RT metal excluding waveguide effect considered no more than 5 m;
- The range of the effective action of the transmitting device must be at least 100 m basing on the fact that the failure of any module would not let self-organizing of radio network (sending GPS signals to other modules);
- The power supply module should be self-contained;

- Construction of receiving and transmitting modules should not create obstacles to the passage in the places of their installation of flaw detection and cleaning projectiles.

The basis (basic instrument) is a standard module "RealEstate 4040R", presented in Figure 8; it has the following main characteristics:

- Pre-amplification factor of AE signals before transmission to air 40 dB;
- The range of the carrier radio frequencies from 30 MHz to 3 GHz;
- Distance of the effective operation of the transmitting apparatus 100 m;
- Stand-alone power supply 3V (two AA alkaline household batteries voltage 1.5 V).

The fully assembled modules go to the installation site.

Installing modules on the dug-in areas made directly before docking pipes in the trench with the installation of APM on the top of the forming inner surface of the pipe wall. Installing modules on the bottom pipe made before docking the ship-handler also preferred setting APM on top of the forming inner surface of the pipe wall.

Case of the MRP sets along the tube axis with the orientation of the front grille of the wind power generator towards to the gas flow. The side edges of the case attached to the metal tube near the corners of welding case cathetus length of 5 mm to 30 ÷ 40 mm. Tips and antennas grab the pipe. Burn-through housing, antenna tips and the pipe wall aren't allowed.

The disadvantage of the system is that the installation of the sensors can be realized only at the stage of construction of the facility, and to exclude the possibility of subsequent prevention.

This limitation does not diminish the relevance of the topic: for example, such a system could equip the Russian section of the bottom pipeline "Turkish Stream", whose length is 910 km from Anapa's Black Sea coast. If the AE signal occurs at this site, tower coastal gas-pumping station (station "Russian") will be informed about it in less than a second.

Thus, a new method of diagnosis gas pipeline by the method of acoustic emission is created. Using a self-organizing wireless network that significantly reduces the time spent on extended scan of the object, does not require stripping of metal and will not yield on the reliability of the existing control of other existing methods, also is one of the most promising Diagnostics of pipelines, which can prevent accidents and reduce their consequences.

References

[1] VSN 012-88 2007 *Construction of Main and Field Pipelines. Quality Control and Acceptance of Work* (Moscow)
[2] STO Gazprom 2-2.4-083-2006 2006 *Instructions nondestructive quality control of welded joints in the construction and repair of commercial and main gas pipelines* (Moscow: LTD "IRC Gazprom")
[3] Kondo K *and* Takada J 1991 *Proc. of Int. Conf. on Motion and Powertransmissions* (Hiroshima: JSME) **2E2** 763–768
[4] Scruby C B 1987 *J. Phys.E: Sci.Instrum.* **20** 946–953
[5] Scruby C B 1985 *Nondestructive Testing* **8**
[6] William B *and* Jolly D 1970 *Materials evaluation* **28 (6)** 135–144
[7] Guo D *and* Mai A 1998 *Review of Progress in Quantitative Nondestructive Evaluation* **17** 485–492

IOP Conf. Series: Materials Science and Engineering **132** (2016) 012016 doi:10.1088/1757-899X/132/1/012016

Temperature Control Method in the Snow Road Construction

Yu Serebrenikova[1], A Lysyannikov[2], Yu Kaizer[2], R Zhelykevich[3], M Plakhotnikova[1], N Lysyannikova[2], M Merko[2] and I Merko[2]
[1]Assistant, Siberian Federal University, Krasnoyarsk, Russia
[2]Associate Professor, Siberian Federal University, Krasnoyarsk, Russia
[3]Professor, Siberian Federal University, Krasnoyarsk, Russia

E-mail: jl.serebro@mail.ru

Abstract. The paper substantiates the process of heat treatment before the snow compaction in snow road construction. The methods to measure the temperature of snow as a moving dispersed material have been considered in the paper.

1. Introduction

Snow roads are constructed to provide transportation network to remote settlements, and to deliver cargo and equipment in severe northern weather conditions. In permafrost areas, snow roads are used during the arctic winter season to provide an improved traffic surface and protect the underlying vegetation and permafrost. Sufficient snow cover and frost depth in the active layer are required to support construction activities.

During the construction of temporary winter roads on virgin snow with frozen ground, it is advisable to use special thermo vibrational snow grooming machines for 1 pass loosening and mixing of snow, warming the snow mass and vibration compaction. To streamline the process for snow road construction, a mobile unit has been designed to construct snow roads or runways; it performs consistently sketch of snow and grinding heat treatment, sealing and coating of snow chains notches. Each of the operations requires a set of equipment including the following devices (figure 1): rotary snow plows 1 equipped with the nozzle 2 for guiding the snowpack to the conveyor 3 and gas burner 4, then through nozzles 5 under the working body to seals 6 (vibro-compaction tool from STM-2), trailed rubber-tired roller 7 and a hard roll with a special profile.

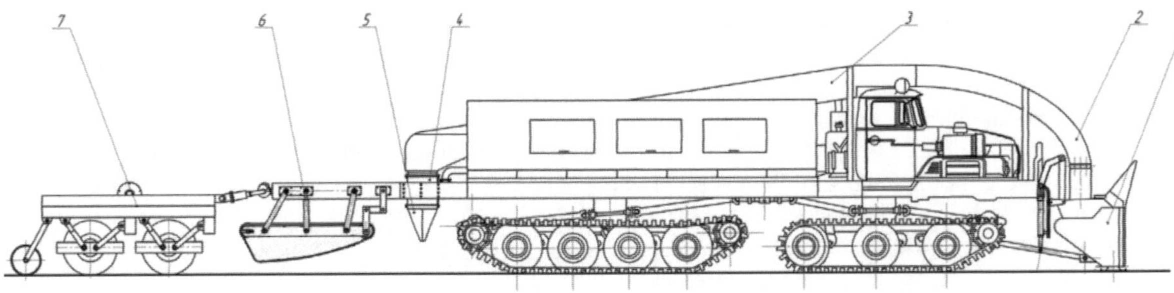

Figure 1. Tracked all-terrain vehicle Ural-5920 with a set of equipment.

According to Construction Codes 137-89 [1], snow road construction requires the snow road density profile to be over 0.65 g/cc. Density and hardness are the most important snow road characteristics. Density depends on the packing efficiency of snow crystals, that is, the degree to which a volume unit of snow is free of void space. Hardness depends on the tendency of neighbouring crystals to bond to each other by ice bridges (sintering). Both density and hardness are metamorphic, that is, they change with the time and surrounding environmental conditions. Methods used in snow road construction alter the state of natural metamorphism to accelerate the rate at which density and hardness increase. These operations will ensure the maintenance quality by maximizing snow grain contacts for optimal sintering and minimizing the labour and equipment operating hours [2–16]. Most experiences in snow road construction result in such values of density and hardness, the preliminary moistening of snow is essential [3].

In order to get a strong road profile, it is important to moisten the snow mass to 10%, when the snow roads are constructed [12]. Moisture of snow depends mainly on the temperature, so to provide the technological process, it is necessary to raise the temperature of snow to -2 … -8^0 C by introducing hot gases, microwave energy, infrared radiation and so on into the snow. The heat treatment intends to raise the temperature of snow immediately before compaction, and provides equalized moisture distribution [13]. The snow is compacted more effectively with the higher snow temperature (figure 2). Strength and durability of the road depend on the temperature.

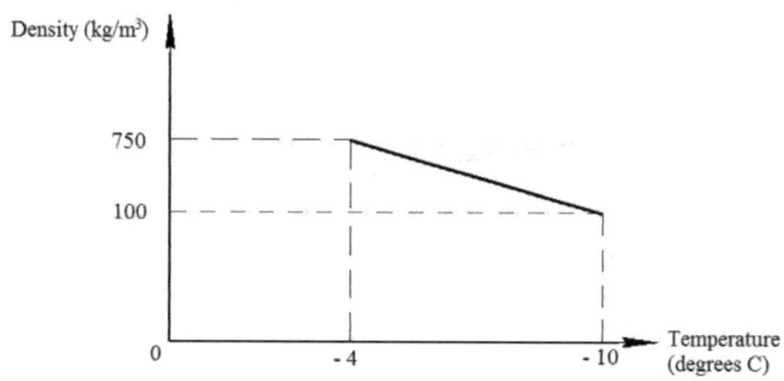

Figure 2. Dependence of the final density of the packed snow on the temperature at the constant maximum load.

The design of a mobile installation involves snow grinding and feeding to the conveyor in the thermal device. Heating of suspended particles in the thermal unit is called a dispersive method of warming, it presents a uniflow convective heat exchanger. Heat exchange occurs in free convection and the heat transfer coefficient is constant. Heat time is equal to the time of snow particles suspension [14]. Therefore, the main task is to maintain snow in the predetermined temperature range to prevent overheating. To do this in the planning phase of the experiment, it is necessary to select a measuring device to monitor the temperature of snow to set the thermal module and to prevent overheating.

2. Instruments and methods for measuring the snow temperature

The temperature control method depends on the range of measured temperatures, precision, operating speed, and the input quantity of heat capacity [15]. The contact method (when the temperature sensor directly contacts with the measured material) is applied to determine low and medium temperatures. The method is based on:

– thermal expansion of liquids, gases and solids (extension thermometers). Thermal expansion is based on the principle of changing the liquid volumes (liquid thermometers) or linear dimensions of solids (deformative thermometers);

– pressure change in closed space with the temperature changes (manometric thermometers). All types of manometric thermometers are almost identical and consist of a thermal bulb, manometric tubular springs (single or multi-turn in the form of bellows) and connecting capillary tube (figure 3);

– change in electrical resistance with the temperature changes (resistance thermometers or thermoelectric thermometers). Measurements with high accuracy are based on the temperature properties of conductors and semiconductors to change electrical resistance when the temperature changes. The higher the temperature, the higher the resistance in conductors. There is an inverse relationship for semiconductors. Quantitatively, the dependence is expressed in the temperature coefficient of electrical resistance. These thermometers consist of a sensor (thermocouples) and protective fittings (figure 4).

– thermoelectric effect (thermocouple).Thermoelectric thermometers are the most widely used devices for measuring temperature because they allow us to define a wide range of measurements, have high accuracy and high sensitivity, allow a remote transmission of readings and automatically record the measured temperature with recording instruments.

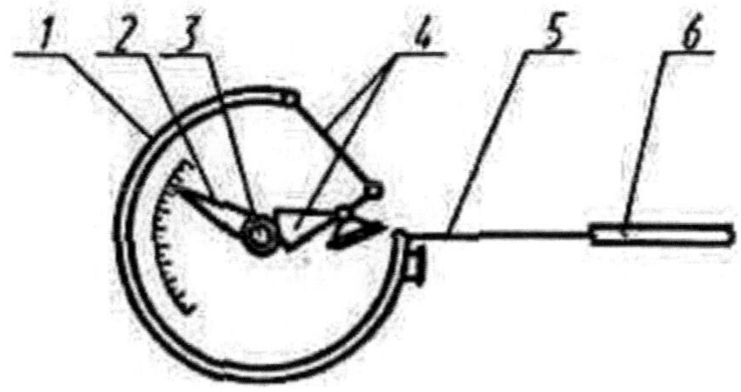

Figure 3. Manometric thermometers: 1 – springs; 2 – an indicating arrow; 3 – an axis; 4 – a gear; 5 – a transmission capillary; 6 – a thermal bulb.

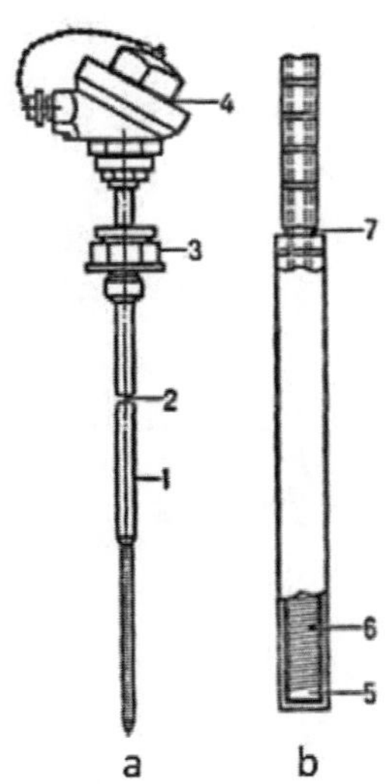

Figure 4. Resistance thermometers: a – general view; b – a sensitive element; 1 – a metal carrying case; 2 – thermocouples; 3 – a pipe connection mounting;
4 –a head for connection to the secondary device; 5 – a micaceous carcass; 6 – winding of platinum wire; 7 – leads.

There are some difficulties in applying contact methods to investigate the snow thermal characteristics:

1. Multiphase systems. A layer of snow is a complex physicochemical system. Its condition depends on the thermodynamic equilibrium of solid, liquid and gaseous components. Solid particles of different shapes and sizes are separated by a gap filled with gas and liquid [16];
2. Measurement occurs at a moving stream of particles in the cross section, at the same time speed and direction of the snow particles have a great importance.

Also, a non-contact method based on the energy of radiation of heated bodies is used to measure the temperature when direct contact with the measured product is impossible. In this case, pyrometers are used. There are three types of the devices depending on the input quantity: radiation, brightness and colour. However, application of these methods to determine the snow temperature is difficult due to low temperatures of snow and complexity of devices design.

However, there is a sensor for continuous indication of the bulk solids operational state (certificate of the authorship SU1656432). The temperature is determined by the contact method: a material stream flowing around the sensor provides virtually instantaneous heating of the thermosensitive plates located on the sensor side (figure 5).

The plates are mounted in thermocouples, which record the steam temperature. Temperature dependence empirically determines the material humidity. The obtained dependence allows us to determine the temperature range in which the moisture content is close to the scheduled one. The fourth thermocouple staked into the base is the control sensor.

The sensor 1 is a hollow case, which is made of thermally conductive material and fixedly mounted in the tray. The front wall 2 is inclined and has the edge 3. The case includes thermally conductive plates 4 with thermocouples 5 and 6 mounted inside. Thermocouples and plates are insulated from each other and from the case. All thermocouples have the output to the recording device.

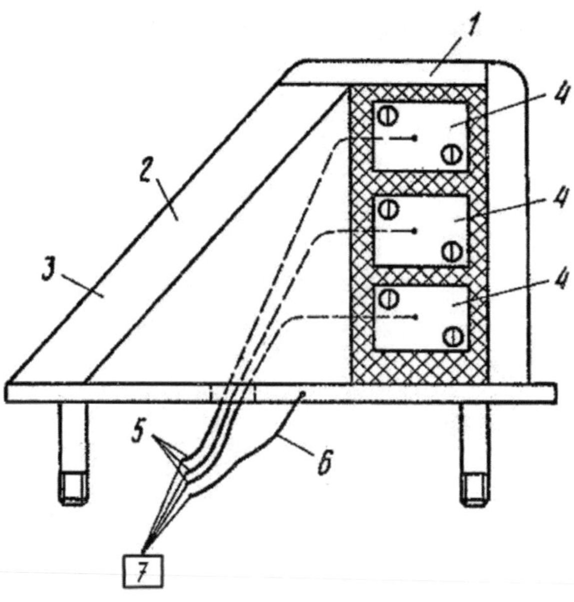

Figure 5. A sensor for continuous indication of the operational moisture state bulk materials: 1 – a sensor; 2 – a heat-conducting wall; 3 – an edge; 4 – thermally conductive plates; 5, 6 – thermocouples; 7 – a recording device.

IOP Conf. Series: Materials Science and Engineering **132** (2016) 012016 doi:10.1088/1757-899X/132/1/012016

3. Summary

The technological process of snow road construction should comply with certain thermal conditions of the snow preliminarily compaction to allow configuration and control of the operating modes of a thermal unit (burner) for heating the snow mass in a predetermined temperature range.

References

[1] VSN 137-89 1990 *Design Construction Maintenance Winter Motor Road* (Moscow: Mintransstroy)

[2] Shoop S A *et al* 2010 *Snow Roads at McMurdo Station, Antarctica* (Hanover, NH: U.S. Army Corps of Engineers) **ERDC/CRREL TR-10-05**

[3] Abele G 1963 *Correlation of Unconfined Compressive Strength And Ram Hardness of Processed Snow* (Hanover, NH: U.S. Army Corps of Engineers) **TR 85**

[4] Alger R G 1993 *The Effect of Grain Size And Compaction on Snow Strength* (Hanover, NH: U.S. Army Corps of Engineers)

[5] Barthelemy J L 1975 *Snow road construction by layered-compaction–Construction and maintenance guide* (Port Hueneme, CA: Naval Facilities Engineering Command) **R 819**

[6] Barthelemy J L 1975 *Snow Road Construction–A Summary of Technology From Past to Present* (Port Hueneme, CA: Naval Facilities Engineering Command) **R 831**

[7] Drope M 1977 *Symp. on Tracks or Wheels* (Calgary, Canadian Society for Terrain Vehicle Systems)

[8] Gow A J *and* Ramsier R O 1964 *Age hardening of snow at the South Pole* (Hanover, NH: U.S. Army Corps of Engineers) **112**

[9] Ramsier R O *and* Sander G W 1965 *Proc. of Int. Symp. on Scientific Aspects of Snow And Ice Avalanches (Davos, Switzerland)* (Gentbrugge, Belgium: Int. Association Scientific Hydrology)

[10] RIF 218-2002 2002 *Snow Construction–General Rules For Design And Construction* (Helsinki, Finland: Association of Finnish Civil Engineers)

[11] Russell-Head D S *and* Budd W F 1989 *Compacted-Snow Runways Guidelines For Their Design And Construction in Antarctica* (Hanover, NH: U.S. Army Corps of Engineers) **89**

[12] Merdanov S M 2010 *Analysis of the snow moisture sources in the snow road construction* (Tyumen)

[13] Serebrenikova J G *et al* 2014 *Proc. of IV Int. Conference "Prospects of development and safety of transport complex"* 307–309

[14] Eresko S P *et al* 2008 *Construction road machines* **3** 40–45

[15] Information on http://allsummary.ru/274-metody-izmereniya-temperatury.html

[16] Pavlov M V *et al* 2013 *Mechanization of Construction* **11 (833)** 34–39

IOP Conf. Series: Materials Science and Engineering **132** (2016) 012017 doi:10.1088/1757-899X/132/1/012017

Simulation of Electrical Characteristics of a Solar Panel

S Obukhov[1], I Plotnikov[1] and M Kryuchkova[2]

[1]Associate Professor, National Research Tomsk Polytechnic University, Tomsk, Russia
[2]Student, National Research Tomsk Polytechnic University, Tomsk, Russia

E-mail: igorplt@tpu.ru

Abstract. The fast-growing photovoltaic system market leads to the necessity of the informed choice of major energy components and optimization of operating conditions in order to improve energy efficiency. Development of mathematical models of the main components of photovoltaic systems to ensure their comprehensive study is an urgent problem of improving and practical using of the technology of electrical energy production. The paper presents a mathematical model of the solar module implemented in the popular software MATLAB/Simulink. Equivalent circuit of the solar cell with a diode parallel without derived resistance is used for modelling. The serie8s resistance of the solar module is calculated by Newton's iterative method using the data of its technical specifications. It ensures high precision of simulation. Model validity was evaluated by the well-known technical characteristics of the module Solarex MSX 60. The calculation results of the experiment showed that the obtained current-voltage and current-watt characteristics of the model are compatible with those of the manufacturer.

1. Introduction

Recent years have seen a rapid growth in the market of solar photovoltaic systems for various practical applications: from small autonomous power systems to large power stations [1].

Modern photovoltaic systems comprise solar panels and a large range of power equipment including various types of electronic power converters to ensure maximum power consumption from the generator, optimum modes of battery charging, electric energy conversion in accordance with the accepted standards, etc.

The right choice of basic energy components and optimization of operating modes are crucial for the efficient use of photovoltaic systems. The main tool to solve the stated problems is simulation, which predetermines the necessity to create universal mathematical models of the basic elements of photovoltaic stations enabling comprehensive study of their operating modes.

The research aims to develop the mathematical model of a solar panel implemented in the software package Matlab/Simulink.

2. Theoretical study

The solar panel is composed of a large number of identical solar cells which are semiconductor photovoltaic cells connected serial/parallel and functioning on the basis of photovoltaic effect. The diode equivalent circuit (figure 1) was used to create the mathematical model of a solar cell, physical processes were described by the following equation (1) [2–4]:

$$I = I_{PH} - I_D = I_{PH} - I_S\left[\exp\left(\frac{q(V+I\,R_s)}{A\,k\,T_C}\right)-1\right]-\frac{V+I\,R_S}{R_{SH}}, \quad (1)$$

generated under solar radiation referred to as photocurrent; I_S is reverse saturation current of the unlit diode (dark current); q is electron charge; R_S is series resistance of the photovoltaic cell; A is the coefficient obtained by comparing theoretical and experimental volt-ampere characteristics of solar cells to characterize the quality of a cell (coefficient takes the values from 1 to 5); k is the Boltzmann constant; T_C is current operating temperature of the device; R_{SH} is shunt resistance of the cell.

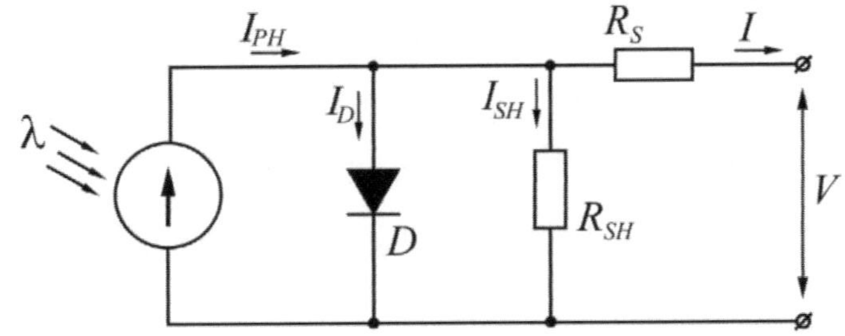

Figure 1. Equivalent circuit of the solar cell.

Equation (1) defines the volt-ampere characteristic of the solar cell to show the relation between output current I and voltage V. The equation is transcendental, contains five variables and has no analytic solution.

To simplify equation (1), make a set of assumptions and use the technical specifications for photovoltaic modules provided by the manufacturer.

The photocurrent value primarily depends on solar insolation λ and operating temperature of the cell T_C:

$$I_{PH} = \left[I_{SC} + K_I\left(T_C - T_R\right)\right]\lambda, \quad (2)$$

where I_{SC} is short circuit current of the photovoltaic cell under standard conditions: temperature of 25°C and solar insolation of 1 kW/m^2; K_I is temperature coefficient of the solar cell short circuit current (from technical specifications); T_R is cell temperature under standard conditions; λ is the value of solar insolation, kW/m^2.

Revers current also depends on the temperature, and it is calculated using the equation:

$$I_S = I_{RS}\left(T_C / T_R\right)^3 \exp\left[q\,E_G\left(1/T_R - 1/T_C\right)/k\,A\right], \quad (3)$$

where I_{RS} is reverse current under standard conditions; E_G is the width of semiconductor band gap (determined by the type of the solar cell).

The value of reverse current I_{RS} can be approximately determined by open-circuit voltage of the element V_{OC} measured at standard temperature of the cell T_R using the expression:

$$I_{RS} = I_{SC} / \left[\exp\left(q\,V_{OC} / k\,A\,T_R\right) - 1\right]. \quad (4)$$

The values of open-circuit voltage and short-circuit current of the solar cell can be determined from technical specifications of the solar panel:

$$V_{OC} = \frac{V_{OC\_R}}{N_S}, \quad I_{SC} = \frac{I_{SC\_R}}{N_P}, \quad (5)$$

IOP Conf. Series: Materials Science and Engineering **132** (2016) 012017 doi:10.1088/1757-899X/132/1/012017

where V_{OC_R}, I_{SC_R} are open-circuit voltage and short-circuit current of the solar panel measured under standard conditions; N_S, N_P are the number of series and parallel elements in the panel, respectively.

The analysis of actual energy characteristics of solar panels proves their sensitivity to minor changes in series resistance R_S, while the value of shunt resistance R_{SH} has no significant impact on their effectiveness [5, 6]. Furthermore, elementary cells of the solar panels are connected in serial chains to obtain the acceptable operating voltage. Therefore, shunt resistance of the panel is basically great and can be neglected without substantial reduction in the model accuracy assuming $R_{SH} = \infty$ [7, 8].

Take the derivative of equation (1) with respect to voltage to determine the value of series resistance R_S:

$$\frac{\mathrm{d}I}{\mathrm{d}V} = -I_S\left[\frac{q}{AkT_C}\left(1+\frac{\mathrm{d}I}{\mathrm{d}V}R_S\right)\exp\left(\frac{q\left(V+IR_S\right)}{AkT_C}\right)\right]. \tag{6}$$

Consider equation (6) with the solar cell in open circuit when V=V_{OC}, I=0. After simple transformations, we obtain:

$$R_S = -\frac{1}{\dfrac{\mathrm{d}I}{\mathrm{d}V}} - \frac{1}{\dfrac{I_S q}{AkT_C}\exp\left(\dfrac{qV_{OC}}{AkT_C}\right)} = -\frac{\mathrm{d}V}{\mathrm{d}I} - \frac{1}{\dfrac{I_S q}{AkT_C}\exp\left(\dfrac{qV_{OC}}{AkT_C}\right)}, \tag{7}$$

where $\mathrm{d}V/\mathrm{d}I=\Delta V/\Delta I$ is cotangent of the slope angle of current voltage characteristic (CVC) of the solar panel at the point of open circuit (easily determined by the experimental CVC from technical specifications).

Equation (1) is solved using Newton's method in which the algorithm to find the numerical solution of equation $f(x) = 0$ is reduced to the following iterative calculation procedure:

$$x_{n+1} = x_n - \frac{f(x_n)}{f'(x_n)}, \tag{8}$$

where $f'(x)$ is the function derivative.

Equation (1) takes the form:

$$f(I) = I_{PH} - I - I_S\left[\exp\left(\frac{q\left(V+IR_S\right)}{AkT_C}\right)-1\right] = 0. \tag{9}$$

With regard to equation (8), we obtain:

$$I_{n+1} = I_n - \frac{I_{PH} - I_n - I_S\left[\exp\left(q\left(V+I_nR_S\right)/\left(AkT_C\right)\right)-1\right]}{-1-I_S\,q\,R_S\exp\left[q\left(V+I_nR_S\right)/\left(AkT_C\right)\right]/\left(AkT_C\right)}. \tag{10}$$

Equations (1)–(10) are the basis for solar panel mathematical modeling.

3. Computer implementation

The current values of solar radiation (G), semiconductor operating temperature (T), and the module parameters determined from the module technical specifications are the initial data to calculate the solar panel characteristics. The function subprogram implementing the solution of the system of equations (1)–(10) is developed in the MATLAB environment to calculate the photovoltaic module parameters. The mathematical model designed as a subsystem is presented in figure 2.

The model is provided with the controllable current source to enable connection of the model with other components of photovoltaic systems. Certain parameters of the solar panel Solarex MSX 60 stated by the manufacturer [9] and presented in table 1 were used to verify the model.

The calculation results have shown that the developed model simulates the current voltage characteristics of a solar panel with satisfactory accuracy: maximum error in the calculated values of short-circuit current and open-circuit voltage within the entire range of possible changes in temperature and illuminance does not exceed 5.4%.

Table 1. Basic parameters of the solar panel Solarex MSX 60.

Parameter	Value
Max (peak) power, P_{max}, W	60
Voltage at max power, V_{max}, V	17.1
Current at max power, I_{max}, A	3.5
Open-circuit voltage, V_{OC}, V	21.1
Short-circuit current, I_{SC}, A	3.8
Open-circuit voltage temperature coefficient, K_U, mV/°C	80±10
Short-circuit current temperature coefficient, K_I, %/°C	0.065±0.015

4. Simulation results

Figure 3 shows the calculated current-voltage and current-watt characteristics of the photovoltaic module at different temperature and illuminance values. The results of calculations confirm the model adequacy in representation of physical processes in the photovoltaic module under various operating conditions.

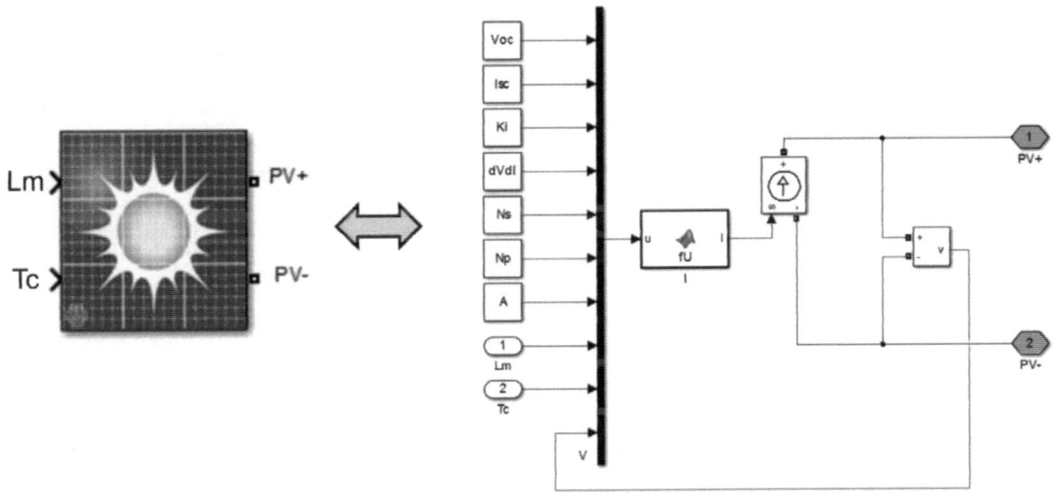

Figure 2. Model of the solar panel implemented in MATLAB/Simulink.

The proposed model can be used to research the operating modes of a single photovoltaic module and its various combinations (solar panels).

5. Conclusion

The model of the solar panel implemented in MATLAB/Simulink has been developed as a result of the research. Certain parameters of the panel Solarex MSX 60 were used to verify the model adequacy. The calculation results showed that the obtained current-voltage and current-watt characteristics of the model are compatible with those of the manufacturer.

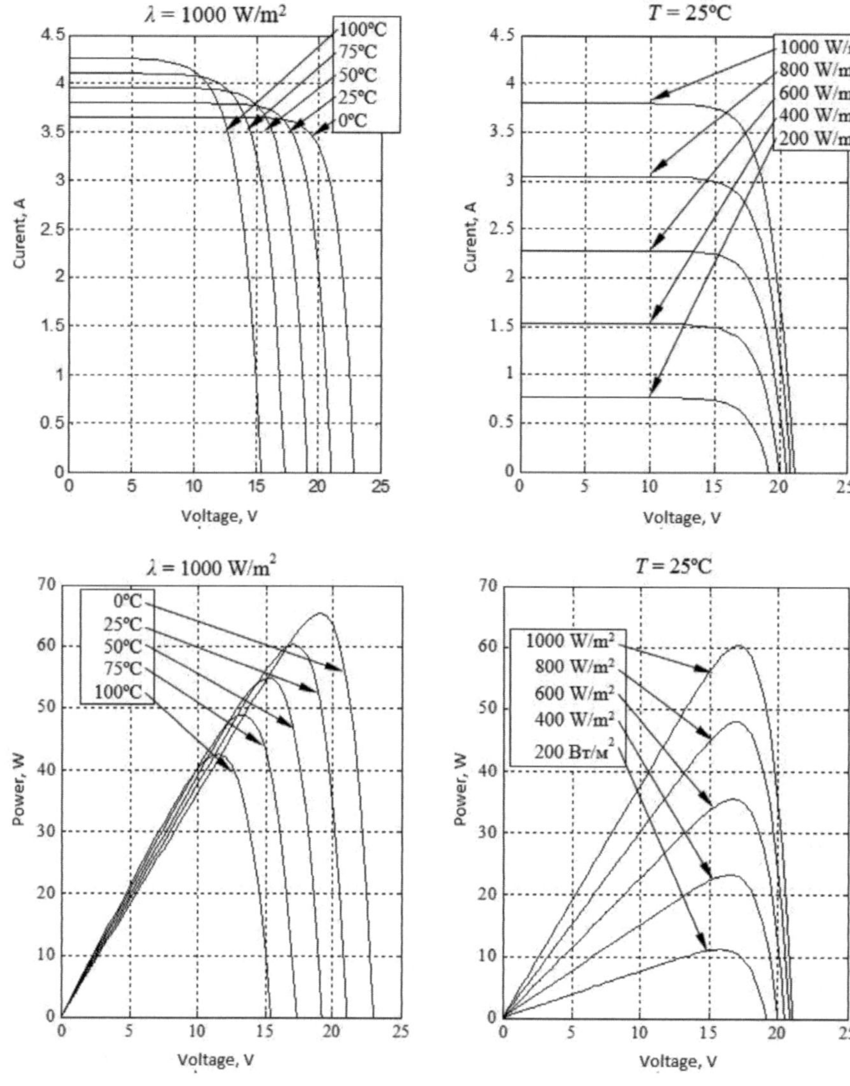

Figure 3. Calculated characteristics of the solar panel Solarex MSX 60.

The model is developed as a separate subsystem capable of the basic parameters input via a dialog box that ensures easy modification and application of the model to study the characteristics of solar cells of different types and to create models of solar panels and photovoltaic systems of any configuration based on these models.

References

[1] REN21 – Renewables 2014 *Global Status Report* URL http://www.ren21.net
[2] Hansen A D *et al* 2000 *Riso National Laboratory, Roskilde* URL http://www.risoe.dk/rispubl/VEA/ris-r-1219.htm
[3] Bonkoungou D *et al* 2013 *Int. J. of Emerging Technology and Advanced Eng.* **3** **(3)** 493–502
[4] Payuk L A *et al* 2016 *Journal of Physics: Conference Series* **671** **(1)** 012044 DOI:10.1088/1742-6596/671/1/012044

[5] Tsai H L *et al* 2008 Development of Generalized Photovoltaic Model Using MATLAB/SIMULINK *Proceedings of WCECS'2008*
[6] Awodugba A O *et al* 2013 *International Journal of Physical Sciences* **8 (22)** 1193–1200
[7] Garg P K *and* Garg V K 2014 *Int. J. of Electronic and Electrical Engineering* **7 (5)** 511–516
[8] Alsadi S *and* Alsayid B 2012 *IJEIT* **2 (6)** 80–85
[9] Information on http://www.troquedeenergia.com/Produtos/LogosModulosSolares/BP-SX-60.pdf

IOP Conf. Series: Materials Science and Engineering **132** (2016) 012018 doi:10.1088/1757-899X/132/1/012018

Creation of Favourable Conditions for Introducing Hydrogen Fuel in Russia

E Boretsky[1] and M Egorova[2]

[1]Student, National Research Tomsk Polytechnic University, Tomsk, Russia
[2]Assistant Professor, National Research Tomsk Polytechnic University, Tomsk, Russia

E-mail: eboretsky@mail.ru

Abstract. The paper presents the key problems that hinder the introduction of new fuel for Russian automobile manufacturers. By analyzing the energy market, as well as trends and forecasts of crude oil and petroleum product prices, it proposes the necessary steps for the Government, along with oil producers and automakers, to gradually convert motor vehicles to hydrogen fuel and, therefore, to occupy a high-potential niche in the energy market. The recommended steps are assumed to be applicable not only to this specific case – introduction of hydrogen to the Russian market – but also to any other alternative fuel in a country with the similar or superior national economy.

1. Introduction

In the second half of the 20th century, people realized that they cause the irreparable harm to the environment. One of the most adverse factors turned to be the harmful substances emitted with exhaust gases of motor vehicles. It became clear that it was necessary to seek an alternative to hydrocarbon fuel.

Many scientists [1] paid their attention to hydrogen which, due to its properties, was recognized as a good substitute for traditional fuels. Having a number of advantages, hydrogen has been already introduced as a supplemental or even primary fuel for some car models. However, hydrogen is an explosive substance, and there is a risk of detonation and explosion when it is stored and transported under pressure. In order to avoid this, it is proposed to use the carbon-hydrogen bond at the molecular level which eliminates the possibility of spontaneous detonation.

The studies present the method of hydrogen accumulation by absorbing it with different carbon modifications [2]. This method provides the opportunity to develop the reusable fuel element with the capacity comparable to the standard fuel tank for hydrocarbon fuel.

Since the creation of the first hydrogen internal combustion engine, many engineering solutions have been proposed to introduce this new energy source. However, none of them have been implemented because of technological or legal barriers.

2. Key preconditions for transition to hydrogen fuel

The current economy largely depends on the energy market. Russia is one of leaders in export of hydrocarbon fuels, and its economic position is very sensitive to price fluctuations of energy products, especially oil. However, the comparison of current oil prices shows that even the most accurate forecasting models may fail [3]. This is why it is impossible to foresee the market trends to the full extent. This is especially the case for long-term trends when oil prices are close to (or below) the critical level for the national economy.

In such circumstances it is necessary to look for new solutions [4] for the Government to significantly reduce the risks of unforeseen events in the global market. The most reasonable solution is to introduce a new alternative product to the global market – such a product that could meet the same demands or even exceed the characteristics of traditional hydrocarbon fuels.

The economic development of the country requires a strong growth of the industry; however, an increased output means increased emissions [5]. But not only the industry and agriculture significantly affect the environment; an increasing number of motor vehicles running on hydrocarbon fuels causes the powerful effects on all biological species of the planet. This is why reduction of emissions generated by exhaust gases of motor vehicles is an important challenge. Use of special filters does not solve this problem [6]. Therefore, the key focus is on the search for the most suitable alternative fuel for motor vehicles.

The most characteristic feature of the 21st century is the replacement of conventional hydrocarbon fuels since hydrocarbon reserves are being rapidly depleted. The current state of the global energy sector is estimated as follows (table 1):

Table 1. The Current State of the Global Economy.

Resource	Share in the global energy mix	Reserves (in raw condition), J
Oil	39%	10^{22}
Gas	23%	10^{22}
Coal	22%	$1.8*10^{23}$
Nuclear energy	7%	10^{30}
Renewable sources	8%	-

The annual energy consumption has reached 10^{20} J and is predicted to grow by 50% by 2020.

At the present time, transport vehicles consume more than a half of petroleum products [5]. Due to the rapid consumption of this fuel, as well as inevitable toxic emissions, the humanity has to find alternative energy sources. The long history shows that hydrogen is the most preferred source of energy.

The advantages of hydrogen include:

1. Being the main source of hydrogen, water is abundant in the Earth. Water reserves are able to meet the growing needs of the humanity for a very long period.

2. Various hydrogen-containing compounds, as well as the huge world's reserves of hydrogen sulfide can be also used as a source of fuel.

3. Hydrogen is a clean energy source, and its addition to a hydrocarbon fuel will reduce toxicity of exhausted gases by 65-75 percent.

4. The use of hydrogen will make it possible to overcome one of the major disadvantages of internal combustion engines – a significant loss of efficiency from 30 to 10 percent at partial loads.

5. It has a high content of energy per unit mass – 120.7 MJ / kg. This is superior to any fossil fuel [7].

The studies in Russia and abroad convincingly demonstrate the cost-effectiveness of hydrogen as a fuel. Figure 1 shows the engine efficiency as a function of the percentage of the air-hydrogen mixture.

It is easy to notice that the greatest efficiency is reached at 100% of the hydrogen content and the air volume of about 2.5 times of the hydrogen volume.

According to this graph, even a minimal addition of hydrogen to gasoline considerably increases the efficiency. The gasoline-hydrogen mixture with 5-10% of hydrogen is proposed to be used for the gradual transition to pure hydrogen. In this case, a volume of harmful emissions will be considerably reduced.

The following conclusions can be made from figure 2.

1. In case of a lean mixture, the CO concentration in exhaust gases decreased to zero and did not depend on a fuel.

2. An increase in the relative concentration of hydrogen in the gasoline-hydrogen mixture significantly reduces emissions of C_nH_m hydrocarbons;

A type of fuel does not affect the generation of nitrogen oxides N_xO_y; however it is determined by combustion temperature of the fuel-air mixture. The use of hydrogen and gasoline-hydrogen mixtures reduces the highest temperature of the cycle inside the internal combustion engine. Hence, it will significantly decrease N_xO_y, concentration and, therefore, volume of emissions [8].

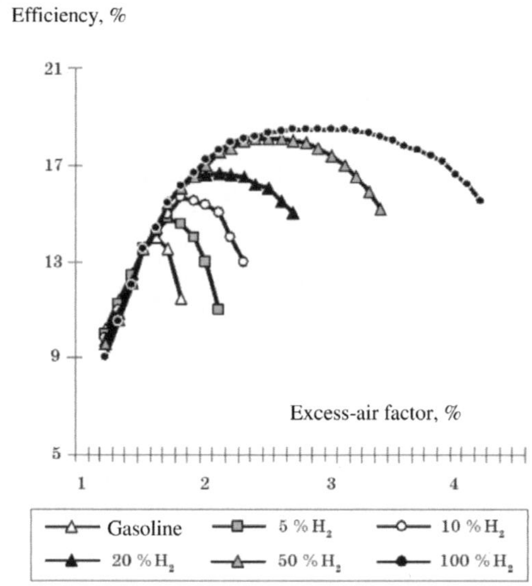

Figure 1. Control performance of the engine as a function of the mixture ratio.

Table 2 shows the performance of harmful emissions during combustion of certain fuels. It is easy to notice that the gasoline has the worst emission rates. The combustion products of hydrogen do not include carbon dioxide and hydrocarbons that affect the environment. Volume of harmful emissions is significantly reduced in the gasoline-hydrogen mixture as well.

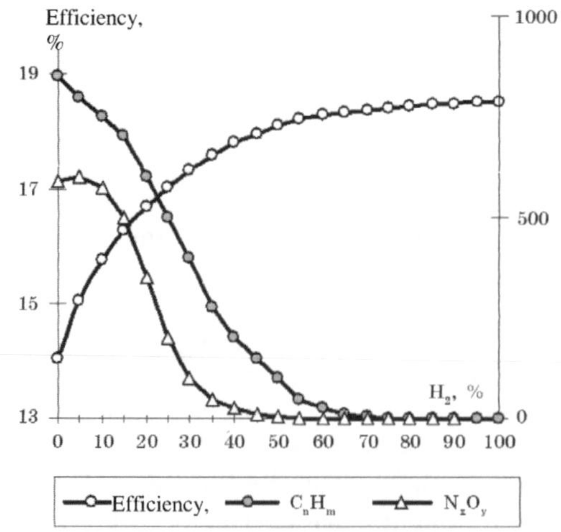

Figure 2. Effects of hydrogen additions on economic and ecological performance of internal combustion engines.

IOP Conf. Series: Materials Science and Engineering **132** (2016) 012018 doi:10.1088/1757-899X/132/1/012018

Nevertheless, with all advantages of hydrogen taken into account, it can be expected that widespread use of hydrogen fuels will be unprofitable to companies engaged in development of oil fields, oil refining and export. Increased use of hydrogen will certainly reduce the consumption of traditional fuels and, therefore, profits of oil producers. In the specific conditions of the Russian society, large companies will likely to prevent any attempt to introduce hydrogen as an alternative to hydrocarbon fuels.

The cost of hydrogen fuel is determined by the relevant infrastructure costs. At the present time, the average cost of industrial hydrogen (99.99% purity) is 61.06 roubles per liter. The average cost of Ai-92 petroleum is 34 roubles per liter. However, it should be taken into account that petroleum has the environmental impact in the form of exhausted gases, while hydrogen fuel does not have such an effect. Although the retail price of hydrogen is significantly higher than that of petroleum, prices of hydrogen fuel are expected to be naturally falling due to the continuous development of technologies and methods of production.

Table 2. Harmful emissions from combustion of different fuels.

Fuel	Emission of harmful substances, g/km		
	CO	CH	NOx
Gasoline	42	8.5	9.1
Liquefied petroleum gas	19	4.8	8.7
Compressed natural gas	8.5	4.5	8.5
Gasoline-hydrogen mixture	3	2.8	4.55
Methanol	28	4.6	4.4
Gasoline-methanol mixture	32	5.4	7.6
Syngas-methanol mixture (H2+CO)	5	2.5	3.5
Syngas (H2+CO)	0	0.4	2.3
Hydrogen	0	0	2.5
EURO 1	2.72	0.93	-

It will be economically profitable and environmentally sustainable to use hydrogen as a fuel (both separately and in combination with petroleum). The multi-year experience of different countries shows that the benefit from the introduction of hydrogen internal combustion engines will be considerable. Nevertheless, this conversion has not been implemented yet for a number of reasons. They include:

1. The lack of clear technical regulatory, administrative and legal framework that would be used to formulate requirements for automakers and fuel corporations. This problem is proposed to be solved by creating the expert committee at the Government of the Russian Federation for coordinating this issue with business stakeholders and research institutes relying on the existing experience in this field.
2. There is no infrastructure for automakers to change the technology and to produce more environmentally friendly cars. The development of the necessary legal framework will provide the possibility to design new and upgrade existing filling stations, as well as to plan logistics solutions.
3. The unsatisfactory economic situation in the Russian automotive industry which suffers significant financial losses from the crisis. The solution of this problem is to provide the governmental support and to attract investments, including foreign ones, for developing new, environmentally friendly models.

This support to automakers must be combined with the strict policy of lean manufacturing in order to identify weak points in production processes and to eliminate or minimize them. It is proposed to use the experience of Toyota in elimination of losses affecting the value of a product. The manufacturing process of Toyota Production System and other companies that have adopted its approach is considered to be consistent with the lean manufacturing concept.

In addition, a large marketing campaign is proposed to restore reputation of Russian automobile manufacturers.

The lean manufacturing policy and the marketing campaign will contribute to production of cars at a lower price and better quality and, therefore, attraction of domestic consumers and investors.

4. Reluctance of domestic consumers to purchase domestic cars. This problem can be solved by improving quality of products. This issue has been detailed in the previous clause.

5. Resistance of oil companies to the increasing demand for hydrogen fuel. It should be reasonably expected that major oil producers will not endure decreasing the demand for their products. They will slow down reduction of the dependence on hydrocarbons by all available means, as this is the case in other countries.

Nevertheless, the gradual replacement of petroleum products as the primary fuel is inevitable. Russian and international corporations will gain significant advantages if they begin to develop the hydrogen fuel market by its own efforts.

3. Measures for Transition to the Alternative Fuel

A number of steps representing the most important operations are suggested for more effective introduction of hydrogen as a new fuel. Implementation of all these steps will ensure the most effective and trouble-free transition not only to hydrogen energy but also to any other alternative fuel.

1. To develop, in cooperation with automakers, the plan for converting motor vehicles to the gasoline–hydrogen mixture. Due to this, automakers will reduce losses from sales of oil products and compensate them by sales of hydrogen.
2. To convert a part of their facilities to production and sales of hydrogen fuel.
3. To create the reliable infrastructure for consumers to assure them that they purchase environmentally friendly cars.
4. Automakers should participate in training of service staff and monitor training of young specialists. To maintain the necessary stock of spare parts.
5. Before putting hydrogen fuel into widespread use, it is necessary to implement a pilot project in one of regions and to demonstrate effectiveness of the new development program to consumers.
6. To significantly expand the production of alternative consumer goods based on petroleum products.

Even today, major oil producers can convert a part of their production facilities and resources to non-fuel products from hydrocarbons. Nowadays, oil is used to produce many everyday products. Almost all polymers serving as raw materials for production of plastics and rubbers, as well as household chemicals such as detergents, coloring materials, solvents, and varnishes are made of oil. We should not forget about the light industry, including synthetic fabrics, fertilizers, and pesticides.

Moreover, oil products are used to produce a wide range of lubricants, including hydraulic and electrical insulating oils, cutting fluids, and petroleum jelly. There are many other products from hydrocarbons, such as coloring agents for artificial limbs, perfume products, strings, paraffin wax for candles and lipsticks, toothpastes, and chewing gums.

Certainly, amount of oil products used for non-fuel production is significantly less than the amount consumed by motor vehicles. Nevertheless, the scientific and technical progress does not stand still: there are increasingly more new substances and materials which may become key oil-based products for the next several decades. This is why R&D costs must be one of the most important elements in cost planning of any oil producer. This will allow them to eventually create new categories of high-demand products.

Thus, these recommended steps will provide the opportunity to gain a share in the hydrogen fuel market and to prepare the industry and the business for the gradual transition to the new source energy. It can be asserted that conversion of production facilities to hydrogen fuel will significantly reduce the harmful environmental impact and the dependence of the Russian economy on world market prices of energy resources.

On the basis of the foregoing, it can be concluded that there is a need for transition to the new fuel, which, thanks to its physical and chemical properties, is capable to eliminate harmful effects on the environment and to significantly improve the technical, economic and operational performance of existing and future internal combustion engines.

The study shows that even a small amount of hydrogen mixed with the gasoline significantly reduces the effect of exhaust gases on the environment, as well as increases engine efficiency and, therefore, reduces fuel consumption.

The trend towards more eco-friendly motor vehicles will lead to the introduction of new energy sources in the coming years. The largest automotive companies are well aware of the current realities and invest significant financial and intellectual resources to find solutions associated with the development of eco-friendly cars. A share of hybrids and models using electricity or hydrogen will only increase, making it necessary to search for new technological solutions.

References

[1] Vinogradov D V 2006 *Problems of atomic science and technology* **1** 153–155
[2] Boretsky E A *et al* 2015 *Russian Physics Journal* **58 (2/2)** 68–72
[3] Jacobson M Z *et al* 2005 *Science* **308** 1901 DOI: 10.1126/science.1109157
[4] Ahluwalia R K *et al* 2004 *J. of Power Sources* **130** 192 DOI: 10.1016/j.jpowsour.2003.12.061
[5] Cheng X *et al* 2007 *J. of Power Sources.* **165(2)** 739 DOI: 10.1016/j.jpowsour.2006.12.012
[6] Ross D K 2006 *Vacuum* **80 (10)** 1084 DOI: 10.1016/j.vacuum.2006.03.030
[7] Jain I P 2009 *Int. J. of Hydrogen Energy* **34 (17)** 7368 DOI: 10.1016/j.ijhydene.2009.05.093
[8] Ramensky A Yu *et al* I 2006 *International Scientific Journal for Alternative Energy and Ecology* **11 (43)** 63–70

Methods of Statistical Control for Groundwater Quality Indicators

E Yankovich[1], O Nevidimova[2] and K Yankovich[3]

[1]Senior Teacher, National Research Tomsk Polytechnic University, Tomsk, Russia
[2]Senior Researcher, Institute of Monitoring of Climatic and Ecological Systems SB RAS, Tomsk, Russia
[3]Student, National Research Tomsk Polytechnic University, Tomsk, Russia

E-mail: yankovich@tpu.ru

Abstract. The article describes the results of conducted groundwater quality control. Controlled quality indicators included the following microelements – barium, manganese, iron, mercury, iodine, chromium, strontium, etc. Quality control charts – X-bar chart and R chart – were built. For the upper and the lower threshold limits, maximum permissible concentration of components in water and the lower limit of their biologically significant concentration, respectively, were selected. The charts analysis has shown that the levels of microelements content in water at the area of study are stable. Most elements in the underground water are contained in concentrations, significant for human organisms consuming the water. For example, such elements as Ba, Mn, Fe have concentrations that exceed maximum permissible levels for drinking water.

1. Introduction

Water quality control is an ongoing important task. Water quality: characteristics of water composition and properties, determining its suitability for specific types of water use. Quality indicators: a list of water properties, which numerical values are compared against the water quality standards. Quality standards: officially established values of water quality indicators for specific types of water use. When organizing the process of water quality assessment, there arises a task of monitoring the groundwater characteristics being within the limits, for the groundwater to be able to meet its intended usage purpose. In the process of monitoring, various parameters of groundwater quality are compared against benchmarks, established in standards, regulations and technical conditions [1].

On the one hand, excessive content of some microelements of natural origin in groundwater may exceed maximum permissible concentrations (MPC), thus making the water unsuitable for consumption by humans, and on the other hand, thanks to the presence of other microelements, drinking water may be more useful for health. To denote concentrations, at which elements entering a human body with water can affect its overall microelement balance, the term of biologically significant concentration (BSC) is used [2]. All microcomponents contained in groundwater can be divided into several groups. The first group – elements that do not produce a substantial impact on the microelement composition of human bodies; the second group – elements contained in biologically significant concentrations (BSC), and the third group – elements, which content in water exceeds the maximum permissible concentrations (MPC) and which are simply dangerous. Maximum permissible concentration of substance in water is such concentration of hazardous substance in water, that does not have direct or indirect impact on human organisms during their entire life and on the health of their future generations, and it does not affect the hygienic conditions of water use [3].

The purpose of the study is to perform control of water quality indicators and to determine the elements contained in underground water in biologically significant concentrations.

Statistical methods allow us to determine, with an accepted level of accuracy and reliability, the studied processes condition in the quality control system. Shewhart control charts are one of the main tools in the arsenal of statistical methods of control [4–5]. Quality control charts, or, shorter, control charts, are used for permanent monitoring of processes, to ensure that the process remains statistically under control from the statistic point of view. Practical application of control charts is mainly confined to using the charts of mean values (X-bar chart) and the charts of ranges (R-chart) [6]. X-bar chart and R-chart are designed to detect the process variations from the average values and the emergence of non-standard ranges in the sample.

2. Research object and method

We have examined the results of chemical analysis of underground water, selected from 343 wells of Ob-Tomsk river value (Russia) in 1962-2009 years. To assess the quality of groundwater, twenty indicators were selected: total iron (Fe), manganese (Mn), fluorine (F), aluminum (Al), arsenic (As), bor (B), barium (BA), strontium (SR), beryllium (Be), chromium (Cr), copper (Cu), cadmium (Cd), cobalt (Co), molybdenum (Mo), nickel (Ni), lead (Pb), mercury (Hg), iodine (I), titanium (Ti), zinc (Zn).

Data on the concentrations of each indicator were collected into samples. A sample includes the array of concentration values of one indicator measured by all wells during a single year. For each sample, the mean value and the range of the sample were calculated. For each indicator, overall average value (mean of the means) was calculated.

To monitor the water quality indicators, X-bar charts and R-charts were built in the "Statistica" system ("Quality Control Charts" module). The horizontal axis on both control charts contains the numbers of the relevant samples (year of sampling); while the vertical axis on the X-bar chart contains mean concentrations of studied indicators, and the vertical axis on the R-chart – the ranges of the relevant samples. Parameters of the charts were defined:

Central line of the X-bar chart corresponds to the mean concentration value for the given indicator, and central line on the R-chart corresponds to the acceptable range of concentration values in the samples.

Threshold limits of the process: MPC - the upper limit of biologically significant concentration, and LLBSC- the lower limit of biologically significant concentration. LLBSC is the amount of a microelement, at which its entry into a human body with drinking water makes 5% of its overall statistical intake [2].

3. Results and considerations

For a continued period of time, in order to conduct water quality analysis, water samples were taken from 343 wells several times a year. The collected water samples were analyzed using standard techniques in accredited laboratories [7].

For each water quality indicator, quality control chars were built. Such charts for some individual indicators are shown in the figures 1–3. The "Quality Control Charts" module of the "Statistica" system was used.

Table 1 contains the quality indicators, number of samples, the average content of microelements in underground water on the studied territory and the relevant standards.

Quality control charts have revealed the difference in behavior of water quality indicators. All average contents of such microelements as aluminum, zinc, cobalt are below the lower limits.

IOP Conf. Series: Materials Science and Engineering **132** (2016) 012019 doi:10.1088/1757-899X/132/1/012019

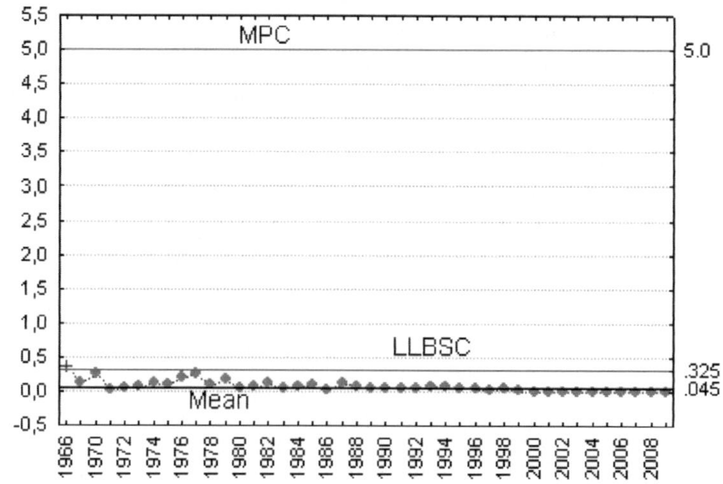

Figure 1. X-bar chart for zinc indicator.

All points of such elements as strontium, chromium and iodine lie within the established borders and their spread from the mean value can be considered as uniform. The substance concentration for them meets the requirements of statistical control for mean level and ranges.

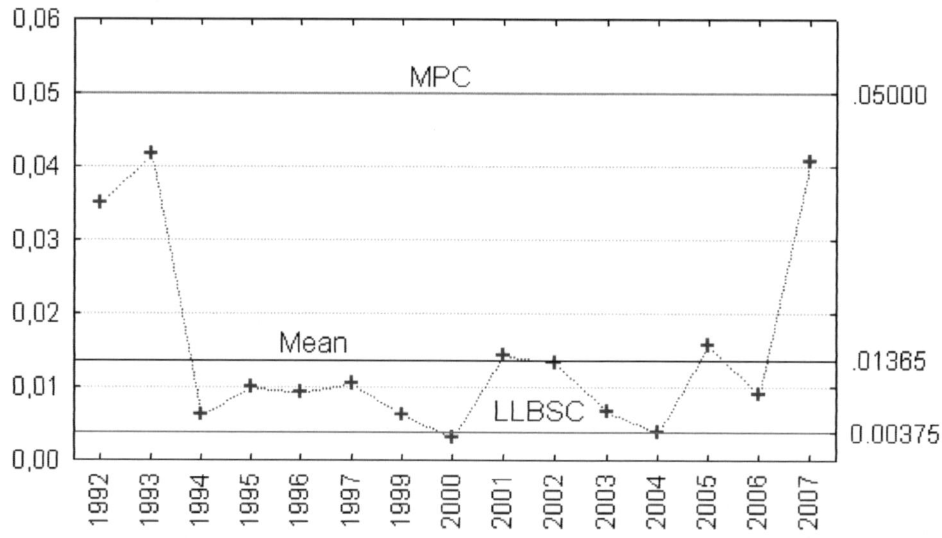

Figure 2. X-bar chart for chromium indicator.

Table 1. Structure of water quality indicators and the standards.

Indicators	Number of samples	Mean, mg/l	LLBSC, mg/l [2]	MPC, mg/l [3]
Al	861	0.14947	0.375	0.5
Ba	656	0.157622	0.02	0.1
Be	158	0.0003	0.00025	0.0002
B	122	0.1137	0.0325	0.5
Fe	10298	5.958	0.375	0.3
I	996	0.21	0.00375	
Cd	262	0.0022	0.0025	0.001
Co	238	0.0032	0.0075	0.1
Mn	5336	0.3980	0.0925	0.1
Cu	4658	0.0497	0.0875	1.0
Mo	412	0.02312	0.00625	0.25
As	530	0.01494	0.00125	0.05
Ni	405	0.0226	0.0075	0.1
Hg	243	0.00045	0.00037	0.0005
Pb	2326	0.01	0.01	0.03
Sr	491	0.51	0.05	7.0
Ti	218	0.07	0.02	
F	4413	0.27	0.05	1.5
Cr	349	0.01392	0.00375	0.05
Zn	4052	0.045	0.325	5.0

All points of average content of iron, manganese and barium are located above the upper limit. This allows us to draw a conclusion that the reason for their increased content is the same and most likely it is related to the geological structure of the territory.

The average contents of lead and mercury relative to the central line (mean of the means) vary on the control charts. In our case, there is a dispersion of mean values; therefore, there are wells with higher concentration levels of separate indicators, compared to other wells.

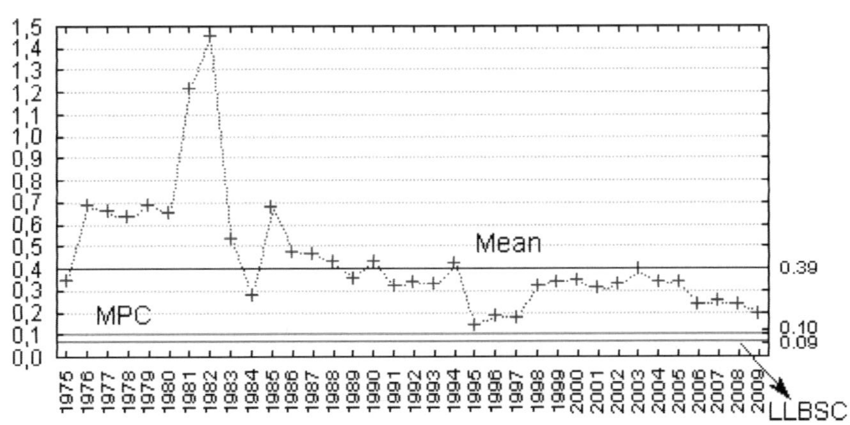

Figure 3. X-bar chart for manganese indicator.

As a result of analysis using the quality control charts, all studied microelements were divided into three groups: Al, Co, Zn, Cu – are contained in concentrations below LLBSC; (B, I, As, Ni, Mo, Hg, Pb, Sr, Ti, F Cr – are contained in concentrations within BSC; Ba, Be, Cd, Mn, Fe – are contained in concentrations above MPC.

Also it is necessary to pay attention to the contents of such microelements as Cu, Mo, Ni, Pb, Hg, as in individual water samples their concentration values exceed maximum permissible concentrations.

4. Summary

The quality control charts analysis has shown that the levels of microelements content in water on the area of study are stable. Fifteen indicators (B, Be, I, As, Ni, Mo, Hg, Pb, Sr, Ti, F, Cr, Ba, Mn, Fe) from twenty are contained in underground water in biologically significant concentrations. Concentrations of five indicators (Ba, Mn, Fe, Be, Cd) exceed maximum permissible concentration (MPC) for drinking water. Such microelements as Al, Co, Zn, Cu are contained in concentrations below LLBSC.

Thus, the method of using the X-bar charts and R-charts allows us to improve quality control of underground water.

References

[1] GOST 27065-86 2003 *Water quality. Terms and definitions* (Moskva: IPK Izdatel'stvo standartov)
[2] Barvish M V *and* Schwartz A A 2000 *Geoecology* **5** 467-473
[3] Hygiene Norms 2.1.5.1315 -03 2003 *Maximum permissible concentrations (MPC) of chemical substances in bodies of water for drinking and community use* (Moscow: The Ministry Of Health Of Russia)
[4] Galtseva O V *et al* 2016 *IOP Conference Series: Materials Science and Engineering* **110** 012094 DOI: 10.1088/1757-899X/110/1/012094
[5] Wheeler D J and Chambers D S 1992 *Understanding Statistical Process Control* (Knoxville, Tennessee: SPC Press)
[6] Montgomery D C 2001 *Introduction to Statistical Quality Control* (N. York John Wiley and Sons)
[7] Yankovich E P *et al* 2015 *Proc. of 15th Int. Multidisciplinary Scientific GeoConf. SGEM* 269–276 DOI: 10.5593/SGEM2015/B12/S2.036

IOP Conf. Series: Materials Science and Engineering **132** (2016) 012020 doi:10.1088/1757-899X/132/1/012020

Possibility of Using Nonmetallic Check Samples to Assess the Sensitivity of Penetrant Testing

N Kalinichenko[1], I Lobanova[2], A Kalinichenko[1], E Loboda[3] and T Jakubec[4,5]

[1]Associate Professor, National Research Tomsk Polytechnic University, Tomsk, Russia
[2]Senior Lecturer, National Research Tomsk Polytechnic University, Tomsk, Russia
[3]Head of Department, National Research Tomsk State University, Tomsk, Russia
[4]Student, Czech Technical University in Prague, Prague, Czech Republic
[5]Student, National Research Tomsk Polytechnic University, Tomsk, Russia

E-mail: nikol_112@mail.ru

Abstract. Versions of check sample manufacturing for penetrant inspection are considered. A statistical analysis of crack width measuring for nonmetallic samples is performed to determine the possibility of their application to assess the penetrant testing sensitivity.

1. Introduction

Liquid penetrant testing (PT) is one of the most widely used nondestructive testing methods for detection of surface discontinuities in nonporous solid materials. It is currently most commonly used surface nondestructive testing method as it can be applied to virtually any magnetic or nonmagnetic material. Penetrant materials and check samples are the tools for PT.

The samples are used to estimate the performance of penetrant materials and to evaluate the sensitivity of the testing techniques. The paper aims to evaluate the discontinuity parameters, such as cracks, in non-metal samples to determine the feasibility of using these samples to assess the sensitivity of the testing technique.

2. Theory

Different check samples containing defects are used to perform penetrant testing. In particular, check samples in the form of plates with transverse cracks dissecting the surface from face to face and those with cracks radiating from the center are widely used. Widespread check samples are samples made of austenitic chromium-nickel steel with star-shaped cracks coated with chromium; samples of stainless steel of the martensitic class with single cracks; brass samples with transverse cracks in the nickel-chrome coating layer. Samples with surface hardened by thermochemical treatment – nitriding, cementation, cyanidation, aluminizing, thermodiffusion chromizing are well-known as well. For all the above check samples, cracks are formed through bending or stretching [1–7]. The authors proposed the techniques for manufacturing check samples of nonmetallic material with preset defect parameters, such as width, depth and length [8].

According to these techniques, crack width is measured by metallographic microscope. The defect dimensions are recorded in the check sample certificate.

Among various problems to be solved using check samples, the most important ones are identification of the penetrant materials efficiency and evaluation of the penetrant testing sensitivity.

The test material is estimated at the beginning of each shift or when changing it, i.e. to inspect a new batch of materials. One or two samples with cracks are used to perform estimation. The sample is tested by penetrant testing using the penetrant materials through the conventional technique. The penetrant material proves to be efficient if the crack is detected. The obtained indication pattern is compared with the flow pattern from the certificate for the check sample. The completeness of crack detection, pattern contrast and its sharpness are considered.

The sensitivity of the procedure is assessed with regard to the preset level of the procedure. Sensitivity is one of the parameters to indicate the technical efficiency of penetrant testing, which characterizes its ability to detect defects. When performing experiments to assess the sensitivity of penetrant testing, the probability of defect detection is taken equal to 0.9 at a confidence level of 0.95 [9].

3. Research

Samples from nonmetallic material manufactured according to the procedure in [8] were used to estimate their applicability for penetrant materials evaluation and assessment of penetrant testing sensitivity.

For the research, the samples were produced with a single crack corresponding to class III sensitivity. The crack width was measured using a measuring microscope MS 50 with a resolution of 1 μm.

According to the method described by Glazkov Yu. A. [9], the techniques used to measure the crack width during sample certification to evaluate the performance of penetrant testing materials are different from those used to evaluate penetrant testing sensitivity.

The measurements taken at 3–5 points along the crack length are sufficient to evaluate the performance of penetrant materials in sample certification. The final measurement result is presented in the form (equation 1)

$$B \pm \Delta b, P \qquad (1)$$

where Δb is the limit of the measurement error, [μm]; P = 0.95 is the established probability of the measurement error within these limits.

The results are presented in table 1.

Table 1. Statistical processing of measurement results to evaluate the performance of penetrant materials on nonmetal check samples.

Sample No.	Width, [μm]				Variation coefficient, [%]
	Average opening width, [μm]	Standard deviation with the establ. probability P = 0.95	Min.	Max.	
11f	18.0	1.55	17.0	21.0	8.00
12f	25.0	1.41	23.0	27.0	5.00
13f	18.5	1.62	16.0	21.0	8.00

A more accurate value of the crack width is to be provided for samples intended to evaluate the threshold of testing sensitivity. Typically, cracks have small fractures, bends and variable width that affect the crack detectability. Therefore, a set of samples to evaluate testing sensitivity should contain samples of the same type with straight cracks.

The disadvantage of metallic samples is changing stresses in the sample which change the depth and width along the crack length. Sample bending under loading results in its curvature, and thus leads to the change in the width along the crack length and crookedness of the obtained defect, as shown in figure 1.

The nonmetal check samples obtained by etching [8] allow implementation of the defects with the desired parameters (opening width, length and depth), which can be seen in figure 2.

Figure 1. Crack in metal check sample.

Figure 2. Crack in nonmetal check sample.

During sample certification, the crack width should be measured at 30–60 random points along the sample length.

The measurement results are statistically processed in order to:
- exclude the known systematic errors from measurement results;
- calculate the measured value;
- determine the RMS deviation of the measurement results;
- check for gross errors and eliminate the errors if needed;
- verify the hypothesis on correspondence of the measurement result to normal distribution;
- calculate the confidence limits of the residual estimation error for the measured value;
- calculate the confidence limits of the estimation error for the measured value [10].

The same samples were used for both the assessment of the penetrant materials performance and evaluation of testing sensitivity. Measurements taken at 50 points along the crack length were followed by statistical processing of the measurement results. The results are presented in table 2.

Table 2. Statistical processing of measurement results for the nonmetal check sample opening width to evaluate penetrant testing sensitivity.

Sample No.	Width, [µm]				Variation coefficient, [%]
	Average opening width, [µm]	Standard deviation with the establ. probability $P = 0.95$	Min.	Max.	
11f	17.0	2.34	12.0	24.0	13.10
12f	24.5	2.22	20.0	30.0	9.07
13f	18.5	2.00	12.0	22.0	11.59

Measurements of opening width of check samples 11f, 12f, 13f presented on figure 3–5.

IOP Conf. Series: Materials Science and Engineering **132** (2016) 012020 doi:10.1088/1757-899X/132/1/012020

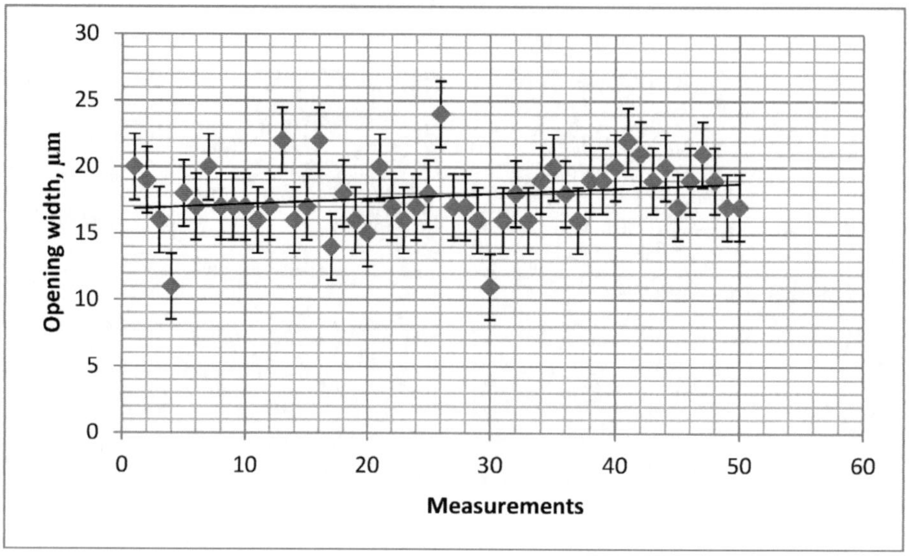

Figure 3. Check sample 11f.

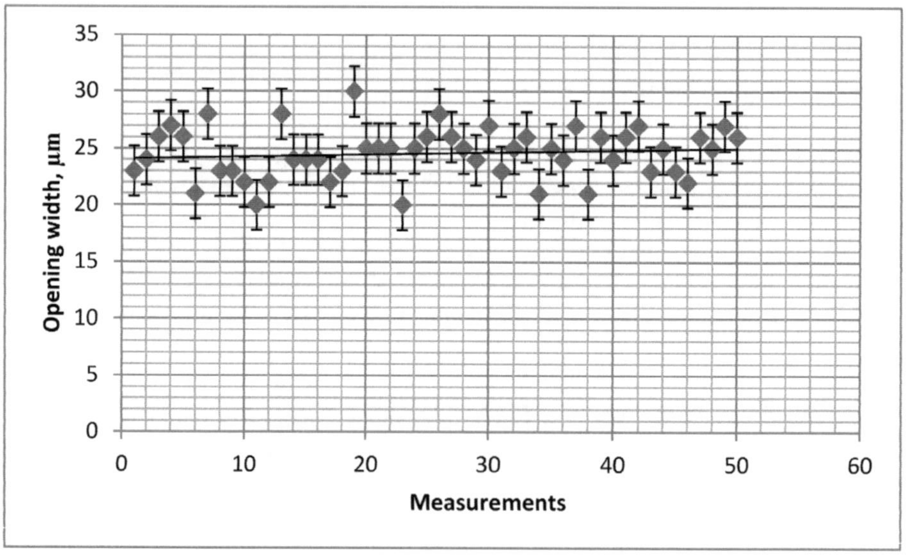

Figure 4. Check sample 12f.

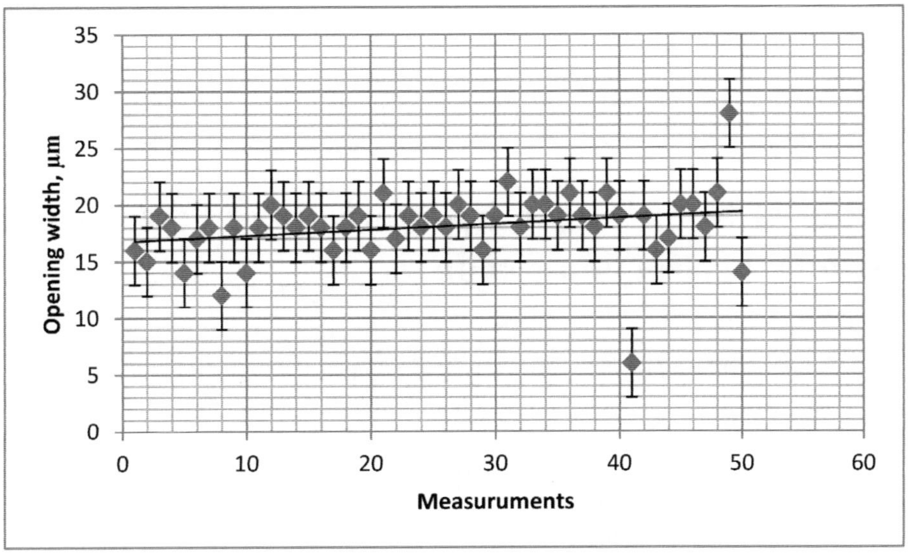

Figure 5. Check sample 13f.

4. Summary

Manufacturing check samples made from metallic and nonmetallic materials used in penetrant testing has been considered.

Metallic samples are less advantageous since cracks vary in their depth and width, and sample bending during manufacturing distorts the crack, which hinders their use for evaluation of penetrant testing sensitivity.

The conducted studies showed that manufacturing the control samples of nonmetallic material through etching provides one-type samples as the width parameters along the crack length are. These samples can be used to estimate both the performance of penetrant materials and to assess penetrant testing sensitivity since the cracks are virtually straight.

References

[1] Alburger J R 1974 U.S. Patent 3 791 198 A
[2] Martin W J 2004 U.S. Patent 6 729 175 B2
[3] Vicki F J *and* Shimizu S 1986 US 4 610 157 A
[4] Dunnwald P 2011 U.S. Patent WO 2 011 097 146 A1
[5] Martin W J 2001 U.S. Patent US 6 311 538 B1
[6] ISO 3452–1:2013 2013 *Non-Destructive Testing Penetrant Testing Part 1: General Principles*
[7] ISO 3452–3:2013 2013 *Non-Destructive Testing Penetrant Testing Part 3: Reference Test Blocks*
[8] Kalinichenko N P et al 2014 *Measurement Techniques* **5** 484 DOI 10.1007/s11018-014-0485-1
[9] Glaskov Yu A 2004 *Defektoskopiya* **9** 74–84
[10] GOST 8.736-2011 2013 *State System For Ensuring the Uniformity of Measurements. Multiple Direct Measurements. Methods of Measurement Results Processing. Main Principles* (Moscow: Standartinform)

IOP Conf. Series: Materials Science and Engineering **132** (2016) 012021 doi:10.1088/1757-899X/132/1/012021

Application of Non-destructive Methods of Stress-strain State at Hazardous Production Facilities

V Shram[1], Ye Kravtsova[1], A Selsky[2], Yu Bezborodov[3], N Lysyannikova[1] and A Lysyannikov[1]

[1]Associate Professor, Siberian Federal University, Krasnoyarsk, Russia
[2]Leading Specialist, NK NOU «NUC «RTS», Krasnoyarsk, Russia
[3]Head of Department, Siberian Federal University, Krasnoyarsk, Russia

E-mail: shram18rus@mail.ru

Abstract. The paper deals with the sources of accidents in distillation columns, on the basis of which the most dangerous defects are detected. The analysis of the currently existing methods of non-destructive testing of the stress-strain state is performed. It is proposed to apply strain and acoustic emission techniques to continuously monitor dangerous objects, which helps prevent the possibility of accidents, as well as reduce the work.

1. Introduction
The Unified System of Conformity Assessment, the current structure of Rostekhnadzor and distillation columns belong to the category of "Equipment for chemical, petrochemical and refining industries with operating pressure up to 16 MPa." According to the rules of Rostekhnadzor, these facilities should be subject to technical diagnosis at least 1 time in 8 years. But first, the rare mode of the survey is unlikely to contribute to the total exclusion of the risk of accidents, secondly, a large size and complexity of the design causes high labor intensity of the full survey by conventional NDT methods.

As a result, gas explosion followed by fire took place in the industrial zone of the Achinsk oil refinery, on June 15, 2014. This incident was caused by the destroyed distillation column. As a result of the technical examination carried out by Rostekhnadzor, it was found that there was depressurization of gas fractionation of the plant columns followed by explosion. The cause of the depressurization was critical thinning of the metal walls of the column supplied to the process piping the hydrocarbon gas as a result of localized corrosion followed by the destruction of the pipeline. Such accidents occurred earlier in Russia (for example, an oil refinery in Kirishi, 2012), and abroad (San Juanito, Mexico, 1984; Texas City, USA, 2005). All these accidents resulted not only in major economic loss, but also loss of life.

Therefore, the search and study of a new method of nondestructive testing of the stress-strain state of distillation columns and piping is an urgent task.

2. Constructive part
Since such accidents are often caused by deterioration of the column adjacent to the technological pipelines, the pre-analysis of the corrosion processes and subsequent destruction was made (Figure 1).

It was found that, due to the impact of condensation, the horizontal sections are most exposed to pitting corrosion, at the upper part of the image, on the sloping surface, erosion grooves can be formed

IOP Conf. Series: Materials Science and Engineering **132** (2016) 012021 doi:10.1088/1757-899X/132/1/012021

at the bottom and vertical sections are weakly subject to deterioration. Gaps in pipes caused by corrosion and deterioration of walls caused by the pressure of the transported medium are usually directed along the generatrix, hence tensile stress acts on the circumference of the cross section [1].

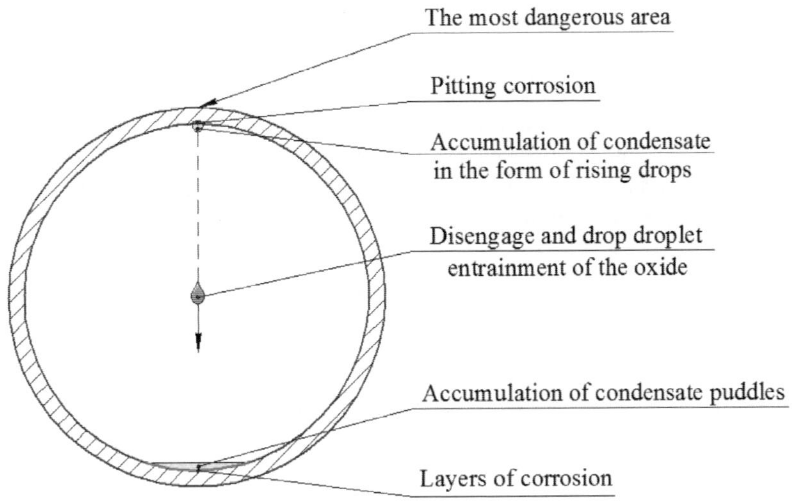

Figure 1. The accumulation of wet condensate in the horizontal section of the pipeline.

We analyzed the existing methods of nondestructive testing of the stress-strain state, their purpose, capabilities and regulatory and technical support (see table 1) [2–6].

Table 1. Nondestructive testing of the stress-strain state of the facilities.

Method	Method	Equipment	Application
Ultrasonic	Normal waves	Special	The extended metal objects
	Body wave	Typical ultrasonic flaw depth gauge with high precision, high-precision thickness measurement	Control efforts fastener tightening compounds
Radiation	Fluoroscopic	Apparatus for fluoroscopy	Laboratory testing of samples
Magnetic	Method of metal magnetic memory (scanning)	Precision coercimeters	Objects of power, steel lifting equipment, steel ropes
	Magnetic noise	Precision coercimeters	No Information
	Magneto-mechanical		No Information
Eddy Current	-	No Information	No Information
Visually-measuring	Visual, measuring	Measuring tool	Laboratory testing of samples
	String	Special	Warp buildings
Optical	Speklinterferometrichesky	Special laser-photographic	Laboratory testing of samples
Strain	Full-scale	Strain amplifier, standard gages	Precise measurement of small deformations of

IOP Conf. Series: Materials Science and Engineering **132** (2016) 012021 doi:10.1088/1757-899X/132/1/012021

			objects
Acoustic emission	Pulse (activity of AE signals)	Universal AE equipment	Voltage detection zones
Thermal (infrared)	Thermal	High-precision thermal imagers	No Information

The results of this analysis showed that strain and acoustic emission techniques are most suitable for this task. The strain technique directly indicates the degree of elastic deformation of the object, while the method of acoustic emission shows the signal activity proportional to the voltage source in the metal (figure 2). These indicators are compared with the elastic limit of steel through the Hooke's law and thus characterize the gravity of the object state [7]:

$$\frac{\Delta L}{L_0} = \frac{\sigma}{E}$$

where ΔL is the amount of the deformation portion (elongation); L_0 is the initial length of the section; σ is tensile stress; E is modulus material.

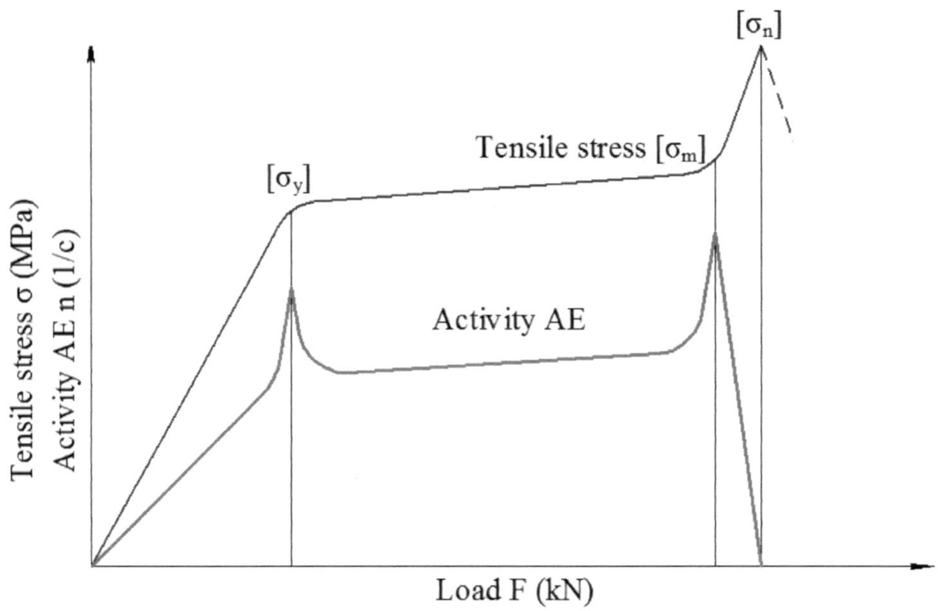

Figure 2. Similarity of the plots of mechanical stress and activity of the AE versus the object load.

Geometrical calculations based on the analysis of the destruction justified the choice of the sensors for these systems: strain is on the top of the generator in the middle of the horizontal portions of the adjacent lines (figure 3); acoustic emission is on the side forming the same areas and a triangulation scheme to scan the body of the column (figure 4).

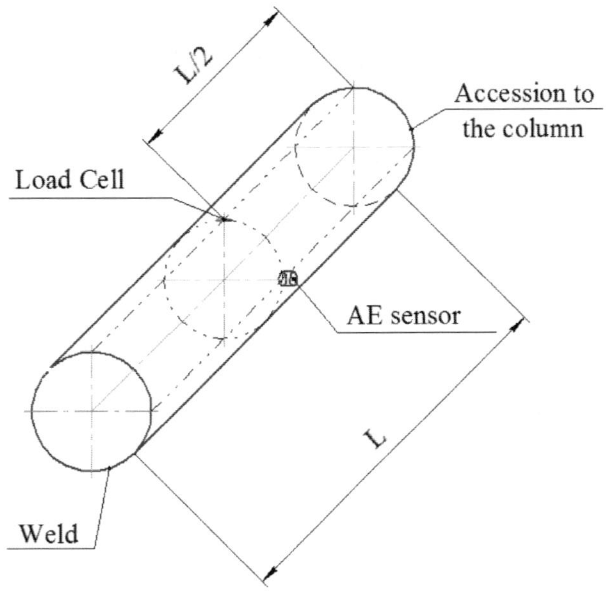

Figure 3. Arrangement of the sensors in the horizontal section of the gas pipeline connected to the column.

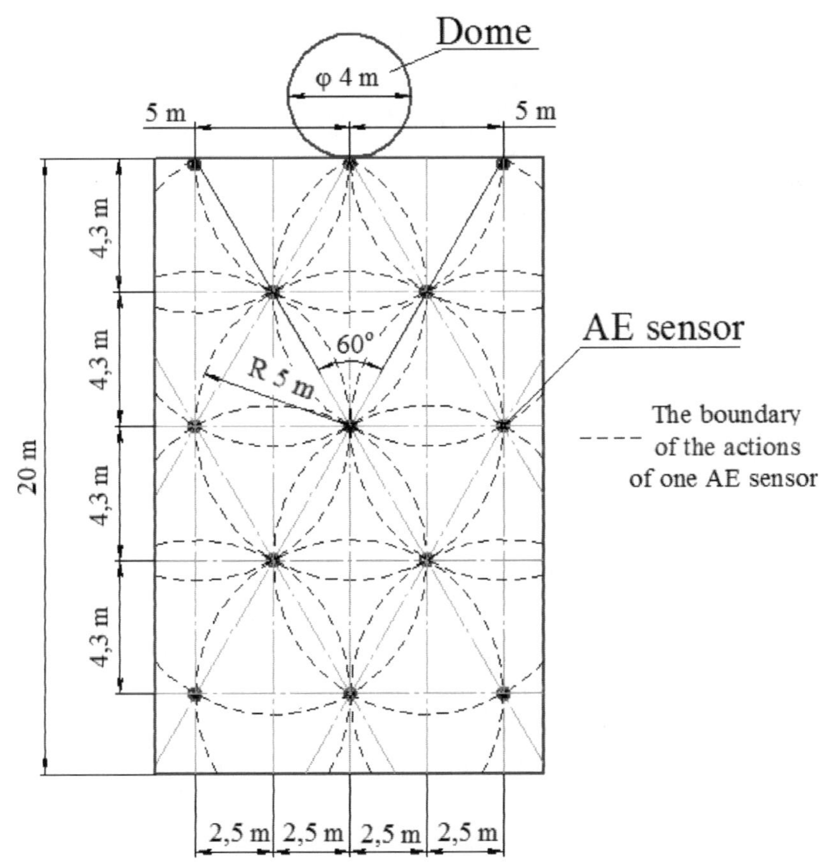

Figure 4. Arrangement of AE sensors to scan the body of the distillation column.

IOP Conf. Series: Materials Science and Engineering **132** (2016) 012021 doi:10.1088/1757-899X/132/1/012021

In order to avoid large amounts of footage and connecting cables wireless sensors to the central system were used. With respect to the AE system, there are such devices in Russia, for example, "EXCITON-4040 P". For the strain gauge system, with the help of specialists in electronics, it was developed the concept of autonomous information transmitters: the amplitude of the broadcast monotone continuous radio frequency is proportional to the degree of the individual tension measuring gage. At the reception center, the frequency of the signal system determines the number and location of the sensor, and the amplitude of the signal, i.e. the degree of deformation of the pipe wall. There is off-grid power for all the sensors from solar panels. Three common rod receiving antenna systems surround the central column equipped with sensors so that it fits perfectly in the triangle made up by them.

3. Summary

Thus, application of the strain and acoustic emission techniques enables continuous monitoring of the object, which helps prevent the possibility of accidents and reduce the complexity of the work performed.

References

[1] Nestleroth J B 2006 *Proc. Of ECNDT (Berlin, Germany)* Mo.2.5.1.
[2] GOST R 53965-2010 2011 *Non-destructive Testing. Evaluation of Mechanical Stresses. General Requirements to the Methods Classification* (Moscow: Standartinform)
[3] Blitz J 1971 *Ultrasonics: Methods And Applications* (Newnes-Butterworth)
[4] IAEA Training Course Series 1992 *Industrial Radiography* (Vienna: IAEA) **3**
[5] IAEA Training Course Series 2000 *Liquid Penetrant and Magnetic Particle Testing at Level 2* (Vienna: IAEA) **11**
[6] Yu T R *et al* 1990 *CIM Bulletin* **83 (937)** 46–52
[7] GOST R 53006-2008 2009 Estimation of potential dangerous objects lifetime on the basis of express methods (Moscow: Standartinform)

IOP Conf. Series: Materials Science and Engineering **132** (2016) 012022 doi:10.1088/1757-899X/132/1/012022

Simulation of the Focal Spot of the Accelerator Bremsstrahlung Radiation

V Sorokin[1] and V Bespalov[2]

[1]Senior Researcher, National Research Tomsk Polytechnic University, Tomsk, Russia
[2]Associate Professor, National Research Tomsk Polytechnic University, Tomsk, Russia

E-mail: sorvb@tpu.ru

Abstract. Testing of thick-walled objects by bremsstrahlung radiation (BR) is primarily performed via high-energy quanta. The testing parameters are specified by the focal spot size of the high-energy bremsstrahlung radiation. In determining the focal spot size, the high-energy BR portion cannot be experimentally separated from the low-energy BR to use high-energy quanta only. The patterns of BR focal spot formation have been investigated via statistical modeling of the radiation transfer in the target material. The distributions of BR quanta emitted by the target for different energies and emission angles under normal distribution of the accelerated electrons bombarding the target have been obtained, and the ratio of the distribution parameters has been determined.

1. Introduction

Bremsstrahlung radiation (BR) sources based on linear electron and betatron accelerators are used to test thick-walled objects [1, 2]. BR generation involves electron acceleration, focusing of the accelerated electron beam towards the target and the interaction of electrons and secondary radiation with the target [3]. When accelerating, the electrons are deflected from the equilibrium orbit straight-line in linear accelerators and circular in betatrons.

The distribution of the accelerated electron current density in the plane normal to the equilibrium orbit is assumed to be normal, which does not contradict theoretical and experimental data. A beam of accelerated electrons is formed towards the target via electromagnetic focusing in linear accelerators or via displacement of the electrons from the equilibrium orbit by the pulsed magnetic field in betatrons. The parameters of the electron current distribution in the beam change, however the distribution pattern remains unchanged. The spectrum of BR emitted by the target is continuous, which is predetermined by the electron energy, the atomic number of the target material and its thickness. As energy increases, the number of quanta in the BR spectrum rapidly decreases. The portion of high energy quanta is relatively small. The resolution of the thick-walled object testing through high-energy quanta is determined by the size of the focal spot of the high-energy portion of the BR spectrum.

The techniques to determine the size of BR focal spot similar to those developed for X-rays [1] are not selective with regard to the energy of the quanta emitted by the target. The prevalence of low-energy quanta determines the results of BR focal spot measuring, but does not provide objective information on the focal spot of the high-energy BR portion. The patterns of BR focal spots formation can be revealed by statistical simulation (Monte Carlo method) of the radiation transfer in the target material. The method enables us to obtain quanta distributions on the target surface for different energies and angles of quanta emission and to analyze them comparing the parameters of distributions of the electrons bombarding the target.

2. Materials and Methods

Distribution of BR quanta emitted by the target is determined by the spatial and energy distribution of the electrons in the beam incident on the target, the atomic number of the target material and its thickness. Typically, the target is made of tungsten or tantalum possessing high thermal properties.

The target thickness-dependent BR emission is extreme. The optimal thickness of the betatron target is typically chosen considering the acceleration energy and by the criterion of the maximum dose. However, the use of target optimum thickness is limited since the targets in serial accelerating chambers are of a specified thickness.

Therefore, the targets of the same thickness are commonly used for different betatron acceleration energies. For example, the tantalum target with a thickness of 0.6 mm is used for an electron energy range of 4–9 MeV. However, for electron beams with high current density which are obtained in linear accelerators, the thickness of the targets is a small part of the electron path in the target material. In this respect, the simulation was performed for tantalum targets of two thicknesses: 0.05 mm and 0.6 mm, the accelerated electron energy being 6 MeV. The PCLab software was used for simulation [4].

The calculation model: a plane-parallel monoenergetic beam with a normal distribution along one of the directions mutually perpendicular and normal to the beam axis (axis X) (figure 1) and uniform distribution along the other direction (axis Y) falls along the normal to the plane target surface, and the number of BR quanta is determined depending on the coordinates of their emission by the target, outgoing direction with respect to the normal to the surface and photon energy.

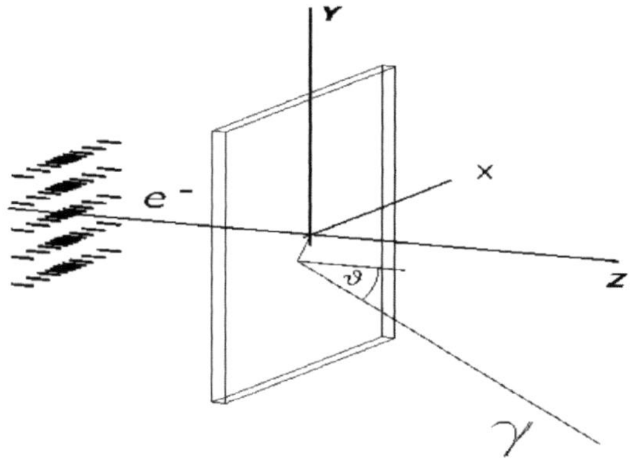

Figure 1. Simulation geometry.

Since the properties of BR by the target exposed to the electron beam with normal distribution are determined by the total contribution of the spot electron beams, the calculated angular and energy distributions of BR for these beams were calculated.

3. Results and Discussion

Increase in the target thickness results in broadening of the BR angular distribution (figure 2) followed by the reduction of the relative contribution of high-energy BR in the spectrum (figure 3). Decrease in the emission angles is followed by the increase in the contribution of high-energy BR. High-energy BR is emitted by the target at small angles to the normal, and as the target thickness reduces, the relative yield of high-energy quanta increases.

The estimated quantum distributions on the target surface are well approximated by the normal distributions for both thin and thick targets (figures 4 and 5) with the dispersion of the thin target equal to the dispersion of the electron distribution in the beam. The coincidence of the spatial quantum distributions on the thin target surface with the normal distribution of the electrons in the beam can be observed in the entire range of the angles of the quantum emission relative to the normal and in all the energy ranges.

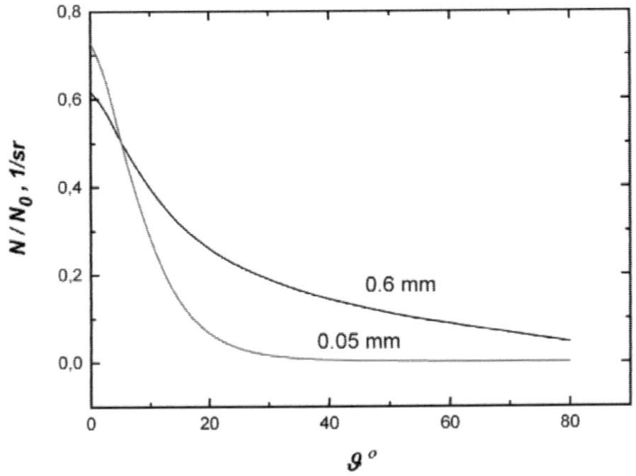

Figure 2. Angular distributions of the BR emitted from the thick and thin targets.

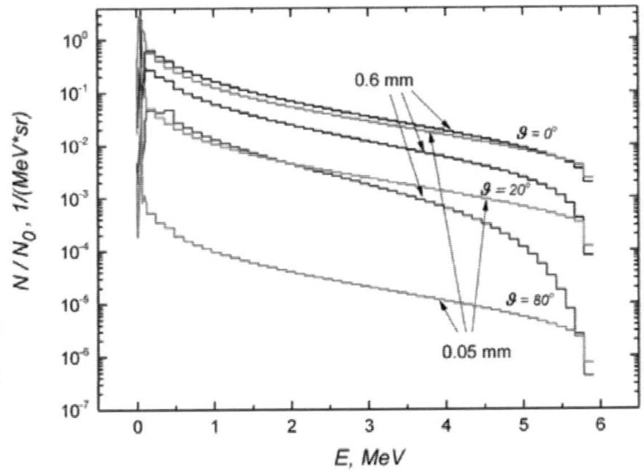

Figure 3. Energy distributions of the BR emitted from the thick and thin targets at angles of 0°, 20°, 80°.

This indicates that the focal spots of all quanta are similar irrespective of the energy and emission angles, and the spatial distribution of BR in the focal spot of the thin target is determined by the spatial distribution of incident electrons only.

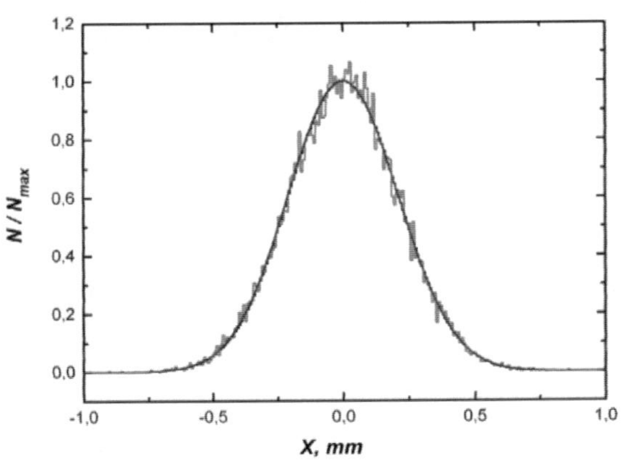

Figure 4. Quantum distribution on a thin target surface (histogram) for normal distribution of electrons in the beam (smooth curve) with a standard deviation of 0.21 mm at the angles of quantum emission by the target relative to the normal within the range of 0–10° and with the energies from 10 keV to 6 MeV.

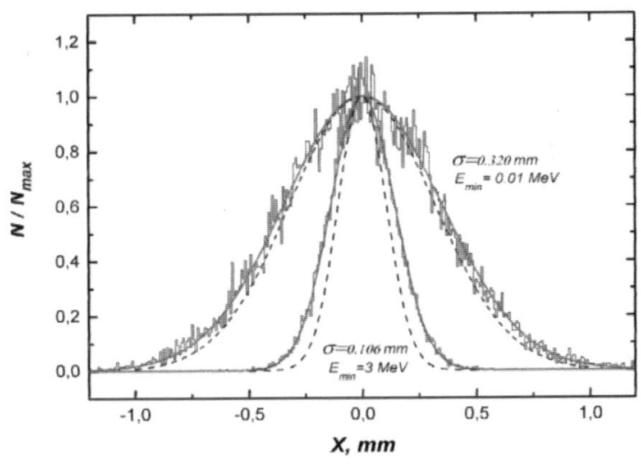

Figure 5. Distributions of quanta with the energies exceeding E_{min} on a thick target surface (histograms and solid curves) for quantum emission by the target at angles within the range of 0–20° under normal distributions of the bombarding electrons (dashed lines) with standard deviations σ.

For the thick target, the quantum distributions on the target surface are different from the electron distributions in the bombarding beam. The distributions depend (figure 5) on both the range of the quantum emission angles relative to the normal and quantum energy, but primarily on the angles of deviation from the normal with narrow distribution of the electrons in the bombarding beam (figure 6).

However, distributions of the quanta emitted by the target along the normal are mainly determined by the bombarding electron distributions. In this case, the smallest deviation of the quantum distributions from the bombarding electron distributions can be observed for the high-energy part of the BR quantum spectrum (figure 7), and the greatest deviations correspond to the narrow distributions of electrons.

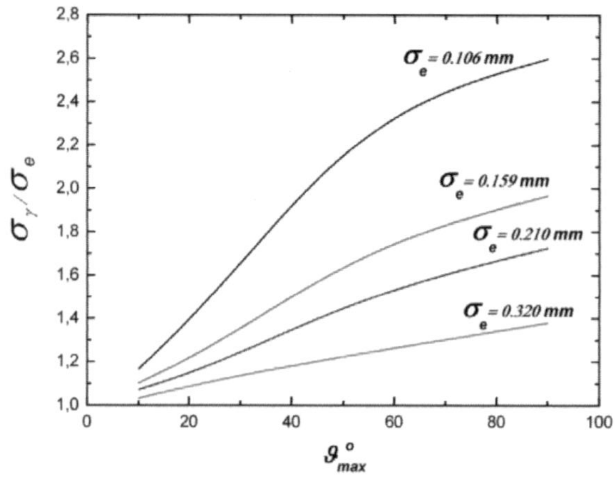

Figure 6. The ratio for standard deviations of quantum distributions σ_γ and bombarding electrons σe depending on the range limit of the emission angles from the target at different σe.

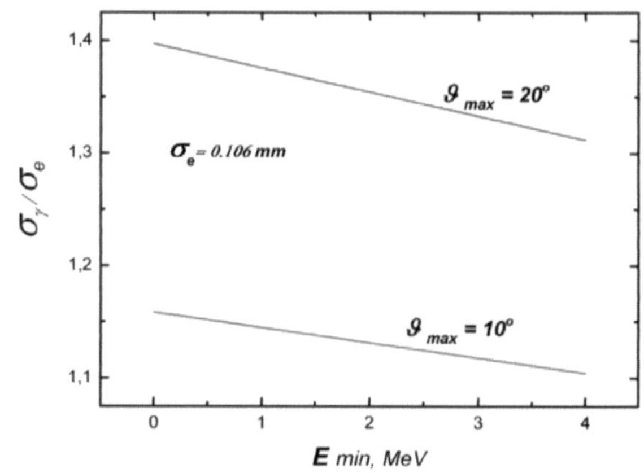

Figure 7. The ratio for standard deviations of distributions of quanta σ_γ emitted by the target at angles ranging from 0 to 10° and from 0 to 20° and bombarding electrons σ_e depending on the range limit value E_{min} of the quantum energy interval ($6 - E_{min}$) MeV.

The properties of the radiation from the target within small deviations from the beam axis are particularly important for practical application. Figures 8–11 show the examples of the characteristics of BR emitted by thin and thick targets within angles of 0–10° under irradiation of the target by electron beam with standard deviation of the normal distribution σe=0.218. The BR spectra (Figure 8) are normalized to the values at energy of 3 MeV. The relative yield of high-energy quanta from the thin target is greater than that from the thick one.

Figure 9 shows the spectra of the BR secondary electrons.

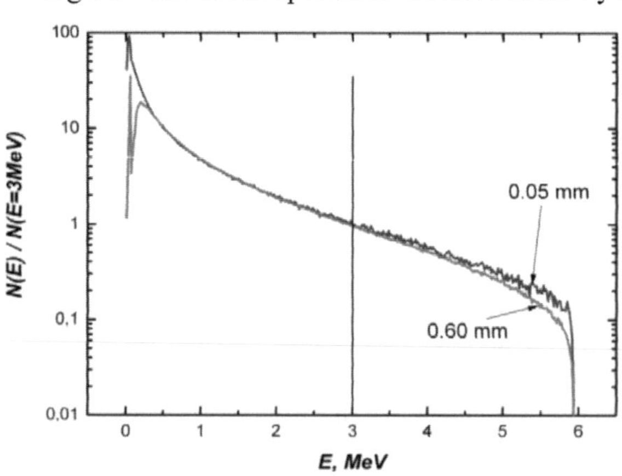

Figure 8. BR spectra emitted by the targets with a thickness of 0.05 mm and 0.6 mm within 0–10°.

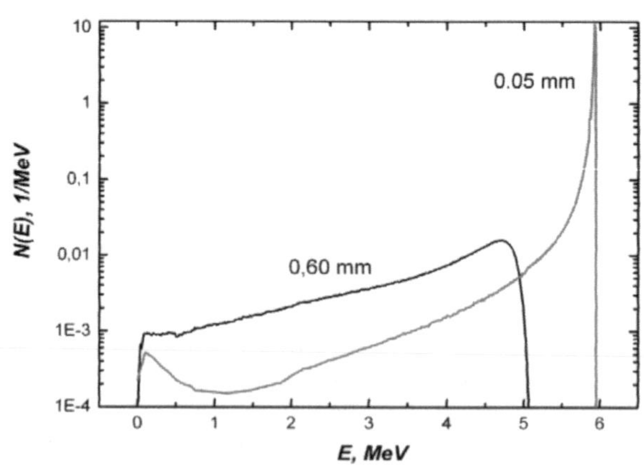

Figure 9. The spectra of the BR secondary electrons emitted by the target with a thickness of 0.05 mm and 0.6 mm within 0–10°.

Angular distributions of BR within 0–10° from the thin and thick targets differ insignificantly (figure 10), and they virtually coincide nearby the beam axis, ϑ=0°. However, the yields of the secondary electrons differ by more than an order of magnitude.

The dispersions of the BR and secondary electron distributions on the surface of the target coincide only for the thin target (figure 11). The dispersion of the distribution of the secondary electrons on the surface of the thick target is much greater than the dispersion of the BR distribution.

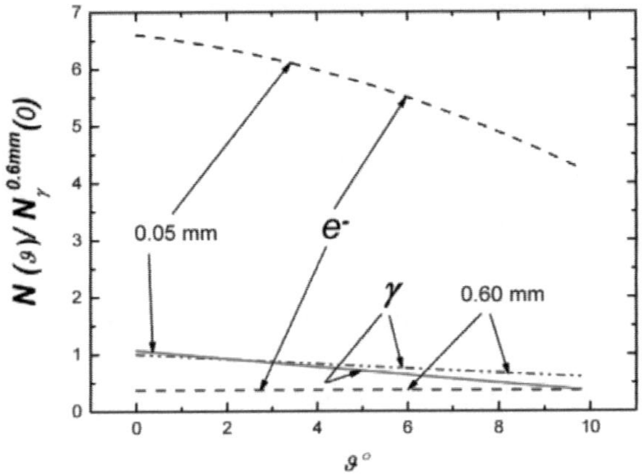

Figure 10. Angular distributions of BR quanta and secondary electrons emitted by the small and large targets within 0–10° normalized to quantum emission by the thick target at an angle of 0°.

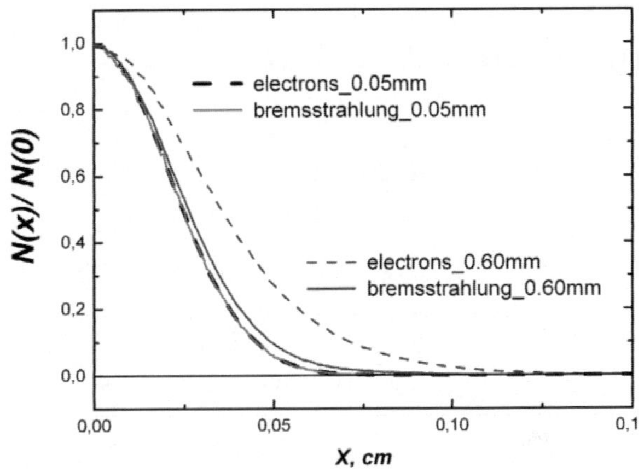

Figure 11. Distributions of BR and secondary electrons on the surface of the thin and thick targets emitted within 0–10°.

4. Conclusion

The size of the BR focal spot is determined by the quantum distribution on the target surface. Simulation of the spatial BR distributions based of the pattern of normal distribution of electrons over the beam cross-section during acceleration, which is approved for accelerating equipment and experimentally confirmed [5], allows evaluation of the possible focal spot oversize in implementing the techniques which disregard the quantum energy.

The obtained results point out the necessity to implement the techniques for accurate focal spot specification because deviations from the optimal implementation conditions can significantly distort the results. The determination of the effective size of the accelerator BR focal spot based on normal electron distribution in the beam bombarding the target is considered to be promising for development of new techniques.

References

[1] Vaynberg E I *et al* 2004 *Proc. Of 16th WCNDT (Montreal, Canada)*
[2] Bjorkholm P *and* Johnson J 2004 *Cargo Security Intelligence* 141–147
[3] Gambaccini M *et al* 2011 *Nuclear Instruments and Methods in Physics Research B*. **269** 1157–1165 DOI: 10.1016/j.nimb.2011.02.089
[4] Bespalov V I 2015 *Software "Komp'uternaya Laboratoriya"* URL http://portal.tpu.ru/SHARED/b/BVI/pclab
[5] Sorokin V B 2015 *Instruments and Experimental Techniques* **58 (3)** 395–398 DOI: 10.1134/S0020441215020256

Development of the Production Process Model Based on the Controllable Input Factors

I Plotnikova[1], L Redko[1], O Titova[2] and O Chaykovskaya[3]

[1]Associate Professor, National Research Tomsk Polytechnic University, Tomsk, Russia
[2]Associate Professor, Siberian State University of Geosystems and Technologies, Novosibirsk, Russia
[3]Professor, National Research Tomsk State University, Tomsk, Russia

E-mail: inna@tpu.ru

Abstract. The research aims to study noncompliances in the production process, defects in chipboard manufacturing and the related costs. The influence of production process parameters on the parameters of finished products and correlation dependence of considered parameters have been shown. The mathematical model of the multiple-factor production process has been developed.

1. Introduction

One of the key factors to ensure the enterprise competitiveness is the quality of products. The product quality is determined not by the quality control and sorting based on the principle of compliance, but mostly by manufacturing environment. The level of noncompliances in the production process output can be reduced by control of the process parameters. The dependence of output parameters of the product quality on variations in the process parameters should be known to reduce noncompliances. To determine such dependencies based on investigation of the process behavior is crucial for enterprises aspiring to the reduction of costs, organization of production process, and enhancement of efficiency of the company's activities and competitiveness as a whole [1].

The combination of factors and conditions of production process results in problems with product quality, but consistent changes in the factors and conditions are capable of eliminating noncompliances. Factors include working facilities; the ability of workers to work in a qualitative and efficient manner; work objects. Conditions include the nature of production process, its intensity, rhythm, duration; climatic conditions in industrial premises; industrial interior and design; the nature of material and moral stimuli; forms of organizing the information service and workplace equipment; social and material conditions for workers [2, 3].

Investigation of influence of the production process parameters on the quality parameters of finished products enables the quality control during the process of manufacturing. This leads to reduction in losses and costs for noncompliances, including sorting reduction, lower cost of the finished product, costs for utilization of unusable products, compensations and payments to consumers due to the possible delivery of defective products.

IOP Conf. Series: Materials Science and Engineering **132** (2016) 012023 doi:10.1088/1757-899X/132/1/012023

The enterprise manufacturing chipboards urgently needs determination of the level of costs due to the occurrence of defective products in the production process, and investigation of types and causes of these costs [4].

The aim of the research was to determine the environment of the chipboard manufacturing process using the method of planning the experiment [5–9] at the stage of the wood chip cover formation. The parameter characterizing the process of production is the chipboard density (kg/m^3). According to the enterprise requirements, the optimum values of density correspond to the limit values of (680–700) (kg/m^3).

2. Research description

Before the development of a mathematical model of the process, a priori information was considered by analyzing the chipboard production process using statistical methods of control. As a result, losses attributable to the issue of defective products were estimated, stages of the chipboard production process with the greatest number of defective boards were defined, types of defects and most common defects were identified in accordance with the rule 80/20. "Short and friable boards" refer to these defects. The causes of their formation in the production process were identified in the study of these types of defects.

The chipboard density (kg/m^3) was defined as the main parameter of chipboard quality. It was revealed that the wood chip cover formation is the stage of the process with the greatest number of defects. The parameters that affect the chipboard quality at this stage and can be regulated during the production process were identified. The parameters are fixed in the process maps by the laboratory technicians every two hours. The parameters are as follows: wood chip moisture, %; resinification of the board inner layer, %; resinification of the board outer layer, %; rate of the wood chip cover formation, mm/s; ratio of layer,%; weight of the wood chip cover, kg; pressing temperature, °C; pressure on the press, kgc/cm^2.

The correlation dependence between the parameters was calculated to reduce the number of considered parameters. If the correlation coefficient (r) evidences the presence of correlated values, one of the values can be excluded from consideration. The evaluation of correlation dependence of the considered production parameters were presented in pairs in the form of scatter diagrams in figure 1.

Figure 1. Correlation dependence between the chipboard inner layer and wood chip moisture.

The absence of tolerable correlation connection indicates that all parameters of the production process should be included in the experiment.

After considering a priori information, the object of consideration was defined in the form of "black box". At this stage, all investigated parameters of the chipboard production process were divided into two groups: factors such as chip moisture, resinification of the inner layer, resinification of the outer layer, and random effects such as rate of the wood chip cover formation, ratio of layers, weight of the cover, pressing temperature, pressure on the press. It was decided to develop a mathematical model for the stage of the wood chip cover formation because the largest qualitative and quantitative amount of defective products is formed at this stage. The quality indicator, namely the board density was chosen as a response.

The next stage of the research is to collect data on the density parameter under different conditions of the production process. The mechanisms and optimum parameters of the chipboard manufacturing can be determined based on these data.

Divide all the parameters into three groups:

- Chip moisture, ratio of layers, pressing temperature.
- Resinification of the inner layer, pressure on the press.
- Resinification of the outer layer, rate of the wood chip cover formation, weight of the cover.

The model for the stage of chipboard manufacturing, that is wood chip cover formation is shown in figure 2.

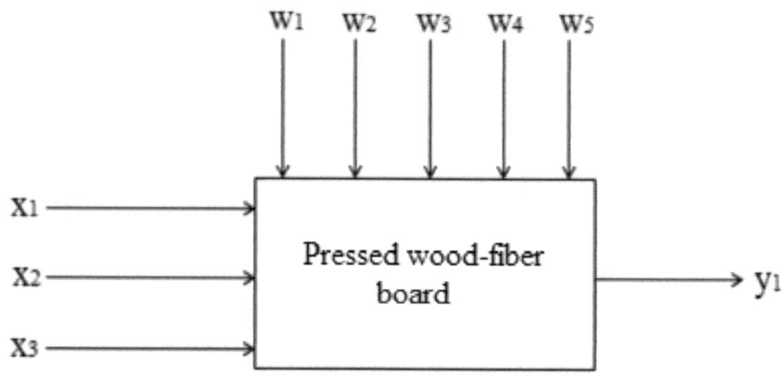

Figure 2. Model of the multi-factor process of chipboard manufacturing.

The factors are designated as: x_1 is chip moisture, %; x_2 is resinification of the inner layer, %; x_3 is resinification of the outer layer, %.

All other parameters correspond to random effects: w_1 is rate of wood chip cover formation, mm/s; w_2 is ratio of layers, %; w_3 is weight of the cover, kg; w_4 is pressing temperature, °C; w_5 is pressure on the press, kgc/cm².

The quality index, namely density y (kg/m³) was taken as a response of the chipboard manufacturing.

The random effects such as rate of the wood chip cover formation, ratio of layers, weight of the cover, pressing temperature, pressure on the press can be varied at this stage. It is impossible to change the factors such as chip moisture, resinification of the inner and outer layers at this stage. They come to the process input with concrete values.

Based on a priori information, the values within which each factor may change its values.

The principle level and variability interval determined for each of the factors are presented in table 1.

All possible factors are determined by the equation 1

$$N = p^k, \quad [1]$$

where p is the number of levels; k is the number of factors.

Each of the eight factors varies at two levels; therefore equation 1 takes the form equation 2:

$$N = 2^k. \quad [2]$$

The factors are divided into three groups; the experiment is to be carried out for each of the groups.

As an example, consider the experimental plan for the first group of factors: chip moisture, ratio of layers, pressing temperature (table 2).

Table 1. Levels of factor variability.

Factors	Factor levels			Variability interval
	Lower level «-1»	Zero level «0»	Upper level «+1»	
x_1	2	2.25	2.5	0.25
x_2	6.7	7.1	7.5	0.4
x_3	10,9	11.55	12.2	0.65
w_1	320	370	420	50
w_2	20	20.55	21.1	0.55
w_3	125	130.5	136	5.5
w_4	195	196.5	198	1.5
w_5	192	221	250	29

Table 2. Extended matrix of planning the experiment of 2^3 type.

x_{01}	x_1	w_2	w_4	$x_1 \cdot w_2$	$w_2 \cdot w_4$	$x_1 \cdot w_4$	$x_1 \cdot w_2 \cdot w_4$
+1	+1	+1	+1	+1	+1	+1	+1
+1	-1	+1	+1	-1	+1	-1	-1
+1	+1	-1	+1	-1	-1	+1	-1
+1	-1	-1	+1	+1	-1	-1	+1
+1	+1	+1	-1	+1	-1	-1	-1
+1	-1	+1	-1	-1	-1	+1	+1
+1	+1	-1	-1	-1	+1	-1	+1
+1	-1	-1	-1	+1	+1	+1	-1

3. Summary

Thus, depending on the input factor value, for instance, chip moisture at the input stage of chipboard manufacturing, namely the formation of wood chip cover, the necessary values of random effects such as correlation of layers and pressing temperature may be chosen so that the chipboard density will correspond to the required values.

References

[1] Deming W E 1986 *Out of the Crisis* (Massachusetts Institute of Technology, Center for Advanced Engineering Study)

[2] Neave H R 1987 *Journal of the Royal Statistical Society. Series D (The Statistician) Special Issue: Industry, Quality and Statistics* **36 (5)** 561570

[3] Evant J R 2005 *Total Quality: Management, Organization and Strategy* (South-Western College Publishing)

[4] Vasendina E A *et al* 2015 *IOP Conference Series: Materials Science and Engineering* **81 (1)** 012070 DOI:10.1088/1757-899X/81/1/012070

[5] Kumje H *et al* 1991 *Statistical method to improve quality* (Moscow)

[6] Montgomery D C 2012 *Reliability and Maintainability Symposium* URL http://www.reliasoft.com/pubs/2012_RAMS_design_of_experiments_and_data_analysis.pdf

[7] Bordunov S V *et al* 2016 *Journal of Physics: Conference Series* **671 (1)** 012009 DOI: 10.1088/1742-6596/671/1/012009

[8] Surzhikov A.P *et al* 2010 *Journal of Thermal Analysis and Calorimetry* **102 (3)** 883–887

[9] Surzhikov A P *et al* 2012 *Journal of Thermal Analysis and Calorimetry* **108 (3)** 1207–1212

Review of Studies of Mechanoelectrical Transformations in Rocks in Russia and Abroad

E Pomishin[1] and L Yavorovich[2]

[1]Student, National Research Tomsk Polytechnic University, Tomsk, Russia
[2]Senior researcher, National Research Tomsk Polytechnic University, Tomsk, Russia

E-mail: lusi@tpu.ru

Abstract. The problem of monitoring and forecast of dynamic manifestations of rock masses becomes immediate in the mining industry because of the growth of mining work intensity and changeover to the mining operations in deeper levels. The article presents a short review of the scientific works of foreign researchers for more complete and in-depth study of geophysical methods of control of the stress-strain state and bump hazard of rock masses.

1. Introduction

Undersurface mine development of solid minerals causes changes in the stress-strain state (SSS) of the rock mass and, as a consequence, geodynamic manifestations. This affects the safety of people, environment, industrial sites and mining operations. Development of changes in the SSS of the rock mass is the result of natural processes in the earth interior and man-made influences. The knowledge of the nature and degree of influence of the elements and processes of technology on the state of the rock masses and its response to this impact becomes especially important. One of the methods used in undersurface mine development of solid minerals is the technological explosion causing the rapid redistribution of the stress field in the rock mass. At this time the stress reduction in one area of the mass is accompanied by the increase in others. As a result of accelerated stress changes in the rock mass in the mine field, bump hazardous situation can occur. In this case, the technology of mining operations is the main man-made factor that provokes the geodynamic events in the areas of critical stress. Unloading of the mass in this way makes it possible to avoid geodynamic events in the field for a while.

Currently, the work on creation of change control information system of the stress-strain state of rock masses and bump hazard forecast is being carried out in the Electronics, Dielectrics and Semiconductors Lab, Tomsk Polytechnic University. The basis of this system is the phenomenon of the dynamoelectric energy transformations in the dielectric structures [1, 2]. Mechanoelectrical transformations are manifested in the form of electromagnetic signals, which can be detected using appropriate instrumentation. Under the laboratory conditions, theoretical and experimental research of the parameters of electromagnetic signals and electromagnetic emission characteristics of model samples and real surface rocks under uniaxial compression and deterministic impulse acoustic treatment are carried out. The stress-strain state change, electromagnetic signal parameters and electromagnetic emission characteristics are logically connected. The research has shown that the mechanoelectrical transformations may occur under thermal, acoustic, mechanical, chemical, radiation and other external influences.

Researches from different countries study geophysical methods of rock-bump hazard evaluation and control of the stress-strain state of rock masses due to the complexity of the object of study and the variety of tasks to be solved in geomechanics and geotechnologies.

2. Review of studies mechanoelectrical transformations

The phenomenon of EME (electromagnetic emissions) from rocks was discovered by J. Milne who reported at the end of the 19th century anomalous magnetic and electric phenomena accompanying an earthquake (Milne, 1890). Intensive research into EME from rocks has been conducted in many research centers since the early 1980s [3].

2.1. Theoretical work

Rabinovitch et al. (2000) attempted to explain the mechanism for EM emissions and concluded that following early pore closure, microcracking and possibly coalescence occurred, while just before peak stress was reached, the rock collapsed. A summary of information about the frequency and wavelength of EM emissions showed that their frequency range was 1 kHz (with a wavelength of 300 km) to 10 MHz (with a wavelength of 30 m) [4].

Goldbaum et al. (2001) identified four distinct EM emissions waveforms: short single pulses, a short chain of single pulses, an extended chain of pulses, and a new group, pulses along baseline voltage changes. Significant to their work were EM frequencies reaching 25 MHz (formerly believed to be only up to 10 MHz) [5].

Rabinovitch et al. (2001) continued investigating mechanisms for EM emissions and concluded that the mechanisms for the earthquake EM emissions were the same as for microfracturing in laboratory tests. They studied the Gutenburg-Richter type and Benioff strain-release relationship for earthquakes and found the relationship extended to the microlevel [6].

2.2. EM Source Mechanisms

In 1986 in the USA Brady and Rowell summarized four mechanisms that cause light to be emitted from fracturing rock: (1) rock fragments frictionally heated to incandescence, (2) electrostatic discharge produced by the deformation of piezoelectric minerals or charge separation on fractured surfaces, (3) plasmas produced by rapid and intense heating of rock material, and (4) excitation of the ambient atmosphere by particle (electrons or positive or negative ions) bombardment. Brady and Rowell concluded that the light emitted from test rocks in the laboratory was caused by excitation of the ambient atmosphere by particle bombardment [7].

In 1988 Chinese researcher, Zi-qiang et al. examined three sources of light: (1) heat radiation from friction, (2) electrostatic discharges produced by piezoelectric effects or charge separation on fractured surfaces, and (3) excitation of the ambient atmosphere by particle bombardment. Because light emissions were observed only at the moment when electrons struck air molecules, the authors concluded that the most likely source of EM emissions was excitation of the ambient atmosphere by particle bombardment [8].

2.3. Laboratory tests

Tuck et al. (1976) tested a cube of quartzite coupled with a quartz crystal to determine piezoelectric emissions when a 0.5-kg hammer was used as a seismic source. They concluded that no piezoelectric fabric was found; therefore, it would be difficult to use EM emissions for the exploration of ore bodies [9].

Nitsan (1977) fractured quartz crystals, tourmaline crystals, and quartz-bearing rocks and recorded EM emissions in the frequency range of 1 to 10 MHz. His interpretation of the source of the emissions was piezoelectricity [10].

Goncharov et al. (1980) tested several large (0.55 by 0.55 by 0.65 m) blocks of concrete containing pieces of granite by applying load and recording both EM and acoustic emissions as the concrete failed. They recognized the fundamental problem of simultaneously recording both EM and seismic emissions and concluded that the number of EM emissions decreased as their amplitude increased. They also found that the ratio of EM to acoustic emissions post-fracturing was 20:1. Prior to fracturing (initial loading), the ratio had been 7:1 [11].

In 1981, Bishop studied piezoelectric effects in quartz-rich rocks. Using a laboratory-designed system, he attempted to prove that the axis of the quartz crystals was a factor in EM emissions. He found that a relationship existed between EM emissions and predictions of the c-axis orientation in quartz crystals [12].

Hanson and Rowell (1982) tested quartzite from the Galena Mine. EM emissions peaked sharply below 40 kHz on three antennas, leading them to conclude that (1) fracture formation coincided with EM emissions, (2) EM emissions fell into a frequency range of less than 40 kHz, (3) EM emissions seemed to be directional, and (4) the amplitude of EM emissions seemed independent of stress, but not independent of stress drop [13].

Khatiashvili (1984) showed that as the size of fractured crystals increased, electrical potential also increased [14].

Zi-qiang et al. (1988) fractured granite in the laboratory and found that the most intensive light pulse and acoustic emissions were recorded simultaneously at the moment of rock fracture [8].

Weimin et al. (1991) fractured quartz, limestone, and granite samples and reported that recorded EM emissions were a result of rock fractures [15].

2.4. Earthquakes

Martner and Sparks (1959) noted electrical potential prior to the arrival of seismic waves at the surface of the ground. About 30 minutes prior to the arrival of main earthquake shocks, Gokhberg and Morgounov (1982) recorded EM emissions at frequencies of 27, 81, and 1.5 kHz and 1.63 MHz. Later, Migunov et al. (1984) documented EM emissions in the frequency range of 0.5 to 50 kHz that were associated with seismicity from earthquakes. Fujinawa and Takahashi (1990) observed EM emissions in the 0.01- to 12-Hz and 1- to 9-kHz frequency bands hours before earthquake activity in Ito, Japan. Fujinawa and Kumagai (1992) observed ultralow-frequency (0.01 to 0.6 Hz) to very low-frequency (1 to 3 kHz) electrical emissions before, during, and after volcanic eruptions [16].

2.5. Underground

Frid et al. (2000) continued their work in the laboratory and attempted to correlate EM emissions with crack dimensions. They found that the amplitudes of EM emissions and their changes with loading were independent of both tensile and shear failure and that they were dependent only on the area of the entire crack [17].

Frid (2001) recognized the value of using EM emission criteria to forecast rockburst hazards in coal mines by using the limiting value of broken coal volume, mine working width, coal seam thickness, and coal elastic properties [18].

Butler et al. (2001) conducted field studies at the Brunswick No. 12 Mine in Canada in an attempt to link EM emissions with seismic activity and also to delineate sulfide ore. They used various antennas covering a range of frequencies from 1 Hz to 4.5 MHz. They found that broadband EM emissions with frequencies up to 800 kHz could be induced by seismicity and blasting. However, results did not confirm that EM emissions preceded seismicity [19].

Vozoff (2002) attempted to demonstrate the use of EM monitoring as a warning system for roof failure in a large coal seam in Australia. He collected three complete datasets and concluded that of the three, one set coincided with a roof fall and was correlated with EM activity, one set might have had a "weak correlation at best," and one set had no EM correlations with roof falls [20].

Research on the use of EME to monitor the level of hazard in headings has been conducted in several centres in China. This method of hazard monitoring has found probably most widespread application in this country – a special system of EME monitoring has been developed there (He et al., 2002). The system has already been applied in nearly thirty mines and the EME method of forecasting rockburst is highly valued since it does not require drilling, does not disturb work in the mine, the equipment is easy to operate and the cost of using the method is low. The system, referred to as KBD5, consists of highsensitivity, wide frequency, directional receiving [3].

3. Conclusion

EME from rocks is a promising precursor to predict hazards in the mine. Further research is needed to precisely determine the phenomenon parameters. The measurements performed on rock samples showed EME to be a good precursor for determining the maximum strength of materials and suitable for determining the state of stress of the rock mass. This has been clearly demonstrated by the laboratory tests and fields studies. So far the systems based on AE have been widely used to determine the fall hazard in mines. But the installation of such systems is time-consuming. The systems based on EME require much less time to install.

References

[1] Bespal'ko A A *et al* 2010 Dynamoelectric energy transfers in a rock mass under explosion load in terms of the Tashtagol mine *J. Min. Sci.* **46 (2)** 136–142

[2] Bespal`ko A A *et al* 2011 Diagnostics of Destruction Zone Development in Rock Specimens During Uniaxial Compression Based on the Spectral Characteristics of Electromagnetic Signal Russian *Russ. J. Nondestruct.* **47 (10)** 41–49

[3] Pralat A *and* Wójtowicz S 2004 Electromagnetic and Acoustic Emission from the Rock – Experimental Measurements *ActaGeodyn. Geomater.* **133** 111–119

[4] Rabinovitch A V *et al* 2000 Fracture Area Calculation from Electromagnetic Radiation and Its Use in Chalk Failure Analysis, International Journal of Rock Mechanics and Mining Sciences. **37** 1149–1154

[5] Goldbaum J *et al* 2001 Electromagnetic Radiation Induced by Percussion Drilling, International Journal of Fracture **111** 15–20

[6] Rabinovitch A *et al* 2001 Gutenberg-Richter-Type Relation for Laboratory Fracture-Induced Electromagnetic Radiation, Physical Review E **65** 4

[7] Brady B T *and* Rowell G A 1986 Laboratory Investigation of the Electrodynamics of Rock Failure *Nature* **321** 1–5

[8] Zi-qiang G *et al* 1988 Light and Acoustic Emission Effects during Rock Fracture *Chinese Journal of Geophysics* **31** 69–75

[9] Tuck G J *et al* 1976 A Search for the Piezoelectric Effect in Quartz-Bearing Rocks *Technophysics* **39** 7–11

[10] Nitsan U 1977 Electromagnetic Emission Accompanying Fracture of Quartz-Bearing Rocks *Geophysical Letters* **4** 333–336

[11] Goncharov A I *et al* 1980 Acoustic emission and electromagnetic radiation at monoaxial compression *Reports of academy of Sciences of the USSR* **255** 821–8241

[12] Bishop J R 1981 Piezoelectric Effects in Quartz-Rich Rocks *Tectonophysics* **77** 297–321

[13] Hanson D R *and* Rowell G A 1982 Electromagnetic Radiation from Rock Failure, *U.S. Bureau of Mines Report of Investigations* **8594** 21

[14] Khatiashvili N G 1984 The Electromagnetic Effect Accompanying the Fracturing of Alkaline-Haloid Crystals and Rock, Izvestiya, Academy of Sciences, USSR. *Physics of the Solid Earth*, **20** 656–661

[15] Weimin X *et al* 1981 Experimental Study of Electromagnetic Emission During Rock Fracture, Advances in Geophysical Research **2** 73–83

[16] Martner S T *and* Sparks N R 1959 The Electroseismic Effect *Geophysics* **24** 297–308

[17] Frid V *et al* 2000 Experimental and Theoretical Investigations of Electromagnetic Radiation Induced by Rock Fracture *Israel Journal of Earth Science* **49** 9–19

[18] Frid V 2001 Calculation of Electromagnetic Radiation Criterion for Rockburst Hazard Forecast in Coal Mines, *Pure and Applied Geophysics*, **158** 931–944

[19] Butler K.E *et al* 2001 *An Experimental Field Trial of Passive Seismoelectric Monitoring* (Canadian Geophysical Union: Canada)
[20] Vozoff K 2001 Electromagnetic Emissions Monitoring to Warn of Wind and Gas Outs, *ACARP Final Report* **25**

IOP Conf. Series: Materials Science and Engineering **132** (2016) 012025 doi:10.1088/1757-899X/132/1/012025

Tool and Method for Testing the Resistance of the Snow Road Cover to Destruction

R Zhelykevich[1], A Lysyannikov[2], Yu Kaiser[2], Yu Serebrenikova[3], N Lysyannikova[2], V Shram[2], Ye Kravtsova[2] and M Plakhotnikova[3]

[1]Professor, Siberian Federal University, Krasnoyarsk, Russia
[2]Associate Professor, Siberian Federal University, Krasnoyarsk, Russia
[3]Assistant, Siberian Federal University, Krasnoyarsk, Russia

E-mail: lysyannikov.alek@mail.ru

Abstract. The paper presents the design of the tool for efficient determination of the hardness of the snow road coating. The tool increases vertical positioning of the rod with the tip through replacement of the rod slide friction of the ball element by roll friction of its outer bearing race in order to enhance the accuracy of determining the hardness of the snow-ice road covering. A special feature of the tool consists in possibility of creating different impact energy by the change of the lifting height of the rod with the tip (indenter) and the exchangeable load mass. This allows the study of the influence of the tip shape and the impact energy on the snow strength parameters in a wide range, extends the scope of application of the durometer and makes possible to determine the strength of snow-ice formations by indenters with various geometrical parameters depending on climatic conditions.

1. Introduction

During rapid reclamation of the northern regions in Russia in order to increase the extent of the hydrocarbonic raw material production, the intensity of traffic (vehicular, aerial) is enhanced, and as a consequence, the transport network is to be developed. Due to remoteness of the reclaimed areas from industrial centers and navigable rivers, poor infrastructure, bad weather and natural conditions (swamps, strong watering of the soil), the most efficient and a cost-effective way to deliver necessary cargoes is roads (snow roads) and airfields with compacted snow cover [1–10].

The snow roads depending on the duration of the season usage could be divided into:

- general, intended for operation only during the period with stable negative air temperatures;
- snow roads with a longer lifetime providing transportation during the winter and part (or all) of the summer.

The requirements to be taken into account to ground the design decisions for snow roads are as follows:

- rapidly constructed from local materials with a maximum construction work mechanization;
- withstand the estimated loads of all types of vehicles (wheeled, caterpillar, sledge train) and to ensure their performance with the estimated speed for the required period of service;
- easy to recover from the destruction and environmental factors [11].

It is necessary to consider that in operation of roads and airfields with a compacted snow surface, in order to ensure the projected the intensity of vehicle traffic and a traffic volume, it is essential to continuously monitor the snow strength.

The tool by Soyuzdornii [11], tool NIAS [12] and tool KISI are used to determine the strength of the snow road cover [13] (figure 1). V.Z. Iofik [14] and a Japanese researcher Kinoshita [15] offer different types and geometric parameters of the indenter in contact with snow.

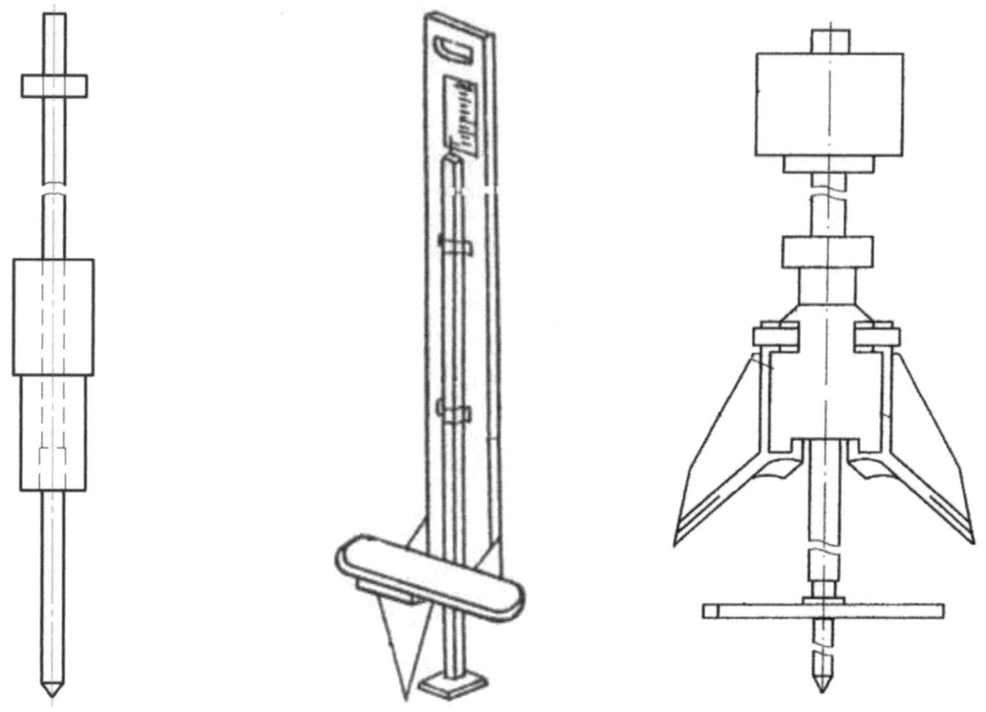

a b c

Figure 1. Tool controls the strength of snow cover: a) tool of Soyuzdornii; b) tool of NIAS; c) tool of KISI.

The main source of error in determining the strength of the snow cover hardness in NIAS is lack of the gauge (it is virtually impossible to put hardness vertically), and the weight determination test is inconvenient when measuring and affect the accuracy of determining the strength of the snow cover. Furthermore, when measuring the strength, the center of gravity is displaced relative to the tester rack, and to balance the longitudinal axis it is necessary to tilt the rack, and along with it, the longitudinal axis of the cone. In this case, the point of the contact of the cone with the surface of the snow cover is broken parallel to the base of the surface cone, wherein part of the energy is wasted on sliding friction of the contact surfaces of the rack with board-focusing and cramps.

2. Results and Discussion

In order to avoid these drawbacks, a device was designed and manufactured (figures 2 and 3). Tool 1 consists of body base 2. The housing is provided with ball bearing 4 (with a split outer race) having a ball head and a through hole in which it is fixed to guide tube 5 installed in the rod with a conical tip 6, 7, exchangeable load 8 and ruler 9 attached to the outer surface of the pipe. In the upper and lower parts of pipe 5 through three equally spaced openings of rectangular diameter 10 are made. Each of the openings 10 is supplied with plug 11 to move longitudinally therein. One end of each plug roller bearing 12 is mounted in contact with the surface of the outer race and the rod being a guide for the movement of the rod along its longitudinal axis. The other end of the plug is made with a thread and is mounted in cylindrical hole 13, and bracket 14 is rigidly secured to guide tube 5. At the end of the threaded plug, nut 15 is installed for adjusting the gap between outer ring rolling bearing 12 and rod 6. To fix the ball head bearing outer split 4, the ball bearing cage is mounted in a slotted body with two arms connected by a stopper. The stopper bolt consists of the brackets and nuts 16 along the longitudinal axis of guide tube 5 and longitudinal groove 17 with transverse shoulder 18. Longitudinal slot 17 includes arrow pointer 19, mounted on rod 6. It indicates the amount of immersion of the conical tip in snow cover.

Figure 2. The appearance of the tool.

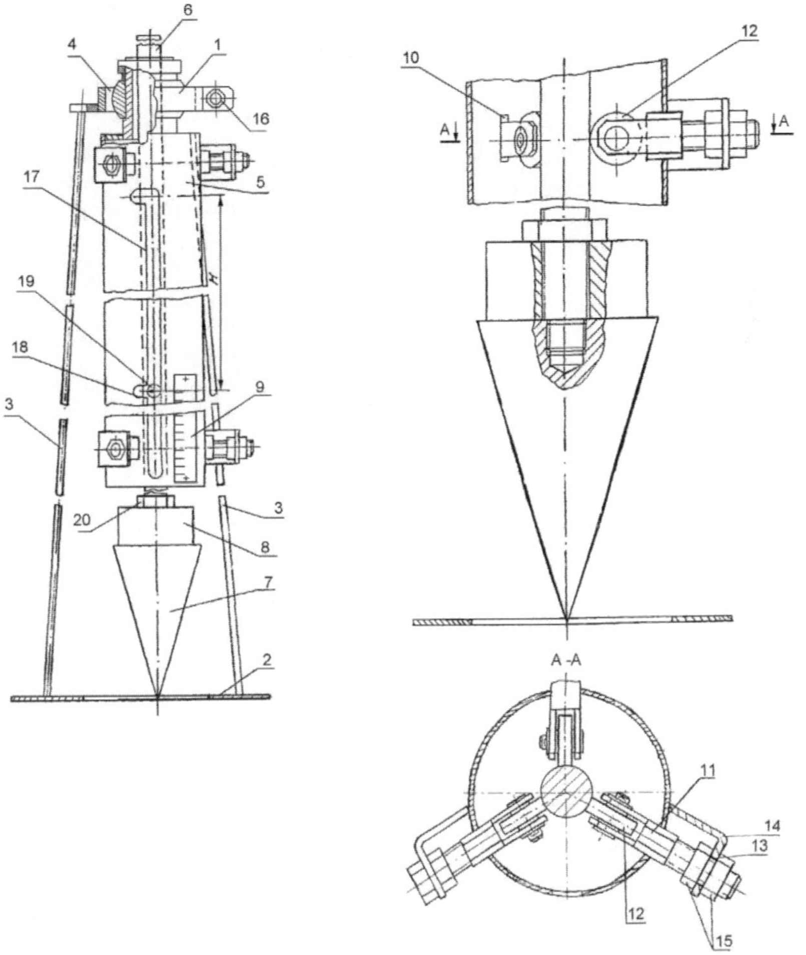

Figure 3. Design of the device 1 – framework; 2 – base; 3 – support; 4 – ball bearings; 5 – the guide tube; 6 – pin; 7 – the conical tip (indentation); 8 – change freight; 9 – ruler; 10 – rectangular opening; 11 – male; 12 – rolling bearings; 13 – cylindrical hole; 14 – bracket; 15 – the adjusting nut; 16 – nut retainer; 17 – longitudinal groove; 18 – cross the ledge; 19 – pointer arrow; 20 – nut.

The tool operates as follows. Framework 1 at bottom base 2 is mounted on a pre-aligned snow cover, with arrow pointer 19 mounted in the lower transverse shoulder 18 of guide tube 5. In this case, the top of conical tip 7 should be in the plane of base 2 in contact with snow, and guide tube 5 under the influence of its weight, the weight of rod 6, conical tip 7, removable load 8 and nut 20 mounted in framework 1 and a holder with vertically split ball. To fix this position, guide tube 5 and framework 1 by rotating nut 16 under the action of the axial force of the bolt stopper contracts the split ball holder that clamps the ball head and keeps it in that position when determining the snow strength. After that, arrow pointer 19 together with rod 6 and conical tip 7 rises up through longitudinal groove 17. The rod is released from this position. Under the action of its own gravity it moves down and in contact with the surface of the conical tip enters into snow. The value of the tip immersion is measured on the scale of the gauge shown in figure 4.

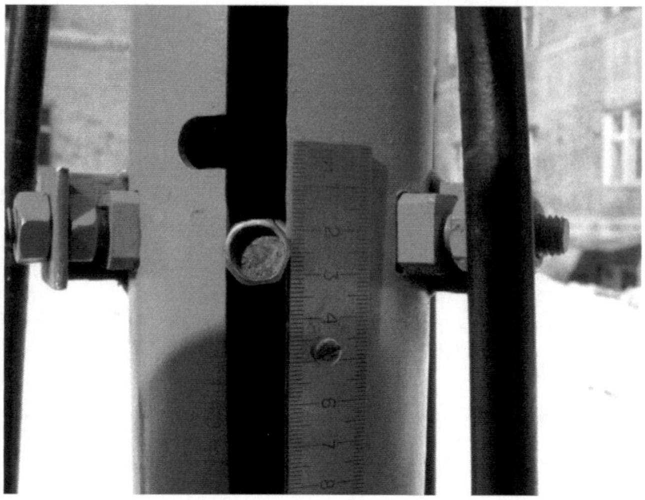

Figure 4. The scale of the ruler.

The strength of the compacted snow depending on the applied load and the depth of the cone is determined by the formula:

$$\sigma = \frac{m \cdot g \cdot H}{\pi \cdot h^2 \cdot tg^2 \cdot \frac{\alpha}{2}}. \quad (1)$$

where σ is snow strength index kPa; m is mass of the rod with a conical tip, [kg]; H is height of the fall of the rod with a conical tip, [m]; h is depth of immersion of the cone, [m]; α is cone angle of 34°12'.

3. Conclusion

The novelty of the proposed technical solution is confirmed by the patent [6]. The feature of the tool is that its design has different impact energy changed by the pitch of fixed lifting rod 6 with a tip 7 (indentation) and changed weight of load 8. This allows the study of the influence of the tip and the energy of the impact on the values of the snow strength in broad limits, extends the scope to determine the strength and hardness of snow and ice formations with indenters of various geometrical parameters depending on climatic conditions. In addition, the use of the device will allow operational services of roads and unpaved airfields to carry out operational testing of the snow compaction degree on the carriageway of roads (winter roads), elements of the airfield grounds and enables efficient control of the parameters of the sealing equipment operation in the construction of winter roads and development of the method of sealing unpaved airfields.

References

[1] Mellor M 1977 *J. Glaciol* **81 (19)** 15–66
[2] Abele G 1990 *Snow Roads and Runways* (Hanover, NH: U.S. Army Corps of Engineers) **90–3**
[3] Adam K M *and* Hernandez H 1977 *Arctic* **30 (1)** 13–27

[4] Alger R G 1993 *The Effect of Grain Size And Compaction on Snow Strength. Contract Report.* (Hanover, NH: U.S. Army Corps of Engineers)
[5] Alger R G 1993 *A New Concept in Trail Grooming. Contract Report* (Lansing, MI: The KRC Groomer, Michigan Department of Natural Resources)
[6] Barber M *and* Brown R L 1993 *Theoretical And Experimental Results of Using Binders For Snow Roads And Runways in Antarctica. Special Report 93–17* (Hanover, NH: U.S. Army Cold Regions Research and Engineering Laboratory)
[7] Barthelemy J L 1975 *Snow Road Construction by Layered-Compaction–Construction And Maintenance Guide. Technical Report R 819* (Port Hueneme, CA: Naval Facilities Engineering Command)
[8] Barthelemy J L 1975 *Snow Road Construction–A Summary of Technology From Past to Present. Technical Report R 831* (Port Hueneme, CA: Naval Facilities Engineering Command)
[9] Diemand D *et al* 1996 *Transportation Research Record: J. of the Transportation Research Board* **1534** 1 DOI 10.3141/1534-01
[10] Drope M 1977 *Proc. of Symp. on Tracks or Wheels (Calgary, Alberta)* (Calgary, Alberta: Canadian Society for Terrain Vehicle Systems)
[11] VSN 137-89 1991 *Design Construction Maintenance Winter Motor Road* (Moscow: Mintransstroy)
[12] Information on URL http://gostrf.com/normadata/1/4293849/4293849566.pdf
[13] Romanov V V 2007 *PhD dissertation* (Technical) Nizhny Novgorod state technical, Nizhny Novgorod, 2007 (in Russian)
[14] Iofik V Z 1990 *The Building and the Road Machines* **5** 31–34
[15] Menis L Y *and* Shipkova M N 1986 *Proc. Of 2nd Int. Symp. (Moscow)*
[16] Zhelukevich R B *et al* 2012 *Bulletin of* Kuzbass State Technical University **2 (90)** 98–100

Control Methods of Operational Properties of Lubricants

N Lysyannikova[1], B Kovalski[2], Yu Bezborodov[3], A Lysyannikov[1], Ye Kravtsova[1], V Shram[1] and M Kovaleva[1]

[1]Associate Professor, Siberian Federal University, Krasnoyarsk, Russia
[2]Professor, Siberian Federal University, Krasnoyarsk, Russia
[3]Head of Department, Siberian Federal University, Krasnoyarsk, Russia

E-mail: nataly.nm@mail.ru

Abstract. Some results of thermal-oxidation and temperature stability testing of motor oils are presented. The catalytic influence of metals on oxidizing processes in lubricants with use of steel 45 was determined. The parameters for identification of oils by groups of operational properties and quantity indicators of the influence of metals on oxidizing processes of lubricants are offered.

1. Introduction

The most urgent problem of modern mechanical engineering is rational application of the lubricants determining durability and non-failure operation of machines. Requirements to lubricants are increasing because it is necessary to provide wear resistance of materials of pair friction in a wide range of loadings, speeds and temperatures.

High temperatures in a combination with active action of oxygen and catalytic action of metal surfaces result in destruction of additives both a base basis and in intensive oxidation of oils with formation of insoluble oxidation products which drop out in a deposit. As a result of oil oxidation, its physical and chemical and operational properties change, viscosity and corrosion activity increases, anti-wear properties worsen.

The temperature has the basic influence on the lifetime of lubricants and causes oxidation and destruction of their base basis and additives. The parameter of thermal-oxidation stability is entered for estimation of anti-oxidizing properties of lubricants which is regulated by specifications and standards on their manufacture on acid number and the sludge formation period [1-3].

A standard method of four-ball lubricant determination of temperature stability with the use of the machine of friction [4] and measurement of the specific size of carbonization lubricants do not provide obtaining objective information.

These methods do not allow investigating lubricant destruction, its intensity and boundary conditions. Therefore, search and substantiation of new methods for research of lubricant temperature stability is a topical problem, solution of this problem will make possible to choose lubricants, to monitor and predict their condition while in service.

The aim of this research is to determine the temperature stability and thermal-oxidation stability of lubricants and substantiation of the criteria for temperature stability and quantity indicators of influence of metals on oxidizing processes of lubricants.

2. Materials and Methods. Experimental Procedure

To study motor oils of a various base basis we have been chosen: mineral Mobil 10W-40 SJ/CH and M-10-$Г_{2к}$; in part synthetic Zic 5000 10W-40 SH/CG-4; synthetic Pentosynth 5W-40 SH/CF.

The tests of motor oils for temperature stability were carried out in a range of temperatures from 140 up to 300 °C in a glass beaker on the device for determination of temperature stability [4]. The tested oil of 50 g was tested within 6 hours, and each subsequent test of the oil test was performed at temperature by 20 °C above the previous one. The temperature of the test was maintained automatically. The method of direct photometric measurement is one of the standard methods for diagnostics of lubricant oils. After testing of each sample the photometer, viscometer and electronic scales were used to determine the factor of absorption of the light flux, viscosity and volatility, respectively. Photometric measurement of oils was carried out for a layer thickness of 2 mm [5].

Thermal-oxidation stability testing of motor oils was carried out for samples of commodity oilM-10-$Г_{2к}$ weighing 100 g. Heating was carried out in a glass beaker, where the oil was mixed up using a glass mixer with a frequency of 300 rev/min. The temperature was set discretely with 10 degree step. Each sample was tested in an interval of temperatures from 150 °C up to 180 °C. After each test, auxiliary devices: a photometer, viscometer and electronic scales to determine the corresponding parameters of optical properties, viscosity and volatility were used. Testing of lubricant oils was carried out up to the value of the light flux absorption equal to 0.7–0.8 units or until the change in the relative viscosity was no more than by 25 %. Comparative evaluation of the changes in the properties of the marketable lubricating oil and the oil and steel was conducted for the factor of the catalytic action [5].

3. Results

Results of the test of the thermal-oxidation stability of oils.

Catalytic and inhibitor influence of metals on oxidation of lubricants was estimated using the obtained experimental dependences. Figure 1 shows graphic dependence of the factor of the light flux absorption on time and temperatures.

At temperatures of 180 °C and 170 °C, insignificant reduction of time of serviceability (curves $1^{/}$and $2^{/}$) of the lubricant oil, as a result of the catalytic actions of the steel 45 sample can be observed. It should be noted that at high temperatures, metal does not have essential influence on the processes a lubricant oxidation, in this case the most adverse factor being the temperature loading destroying additives and the basis of the mineral motor oil. This means that temperatures 180 °C and 170 °C are critical for the given mark of the motor oil.

At temperature 160 °C (curves 3 and $3^{/}$), we can observe an inhibitor action of steel 45 caused by formation of protective chemisorption layers on the sample surface which perform a protective function and interfering increase of speed of chemical transformations in oil. At temperature 150 °C (curves 4 and $4^{/}$), a reverse tendency is observed. Hence, at lower temperatures, chemical reactions of formation of a blanket on the surface of the solid body proceed very slowly, therefore, metal provides catalytic action on the lubricant. At high temperatures, these processes are accelerated, which is characterized by a dark blanket on the sample surface.

The obtained graphic dependences of the factor of the light flux absorption K_a on time and temperatures are described by the polynom of the following kind:

$$K_a = At^3 + Bt^2 + Ct + D, \quad (1)$$

where A and B are the factors describing intensity of formation of oxidation products; C is the factor dependent on the base basis of the lubricant and quality of additives; D is the factor describing the initial optical properties of oil; t is time of the test.

Influence of temperature and steel 45 on change of viscosity during thermal-oxidation stability tests o is expressed by the factor of the relative viscosity K_μ, and the determined relation is [5]:

$$K_\mu = \mu_0 / \mu_{ini} \quad (2)$$

where μ_o is viscosity of the tested oil, cS; μ_{ini} is initial viscosity (commodity oil), cS.

The dependence of the relative viscosity on time and temperature of test $K_\mu = f$ (t) is shown in figure 2. The viscosity of the lubricant in many respects depends on modes of operation and on the degree of its contamination by insoluble products of oxidation, and also on its dispersing-stabilizing properties.

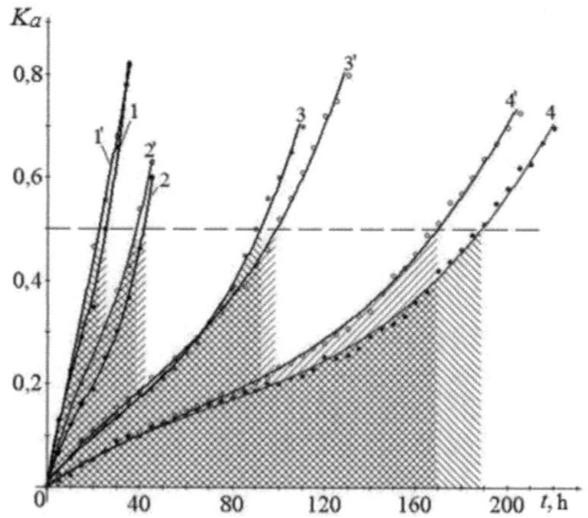

Figure 1. Dependence of light flux absorption factor K_a on temperature and time of the motor oil test M10-Г$_2$к: 1-4 - commodity oil at temperatures 180°C, 170°C, 160°C, 150°C; $1^/$–$4^/$, respectively, for steel 45.

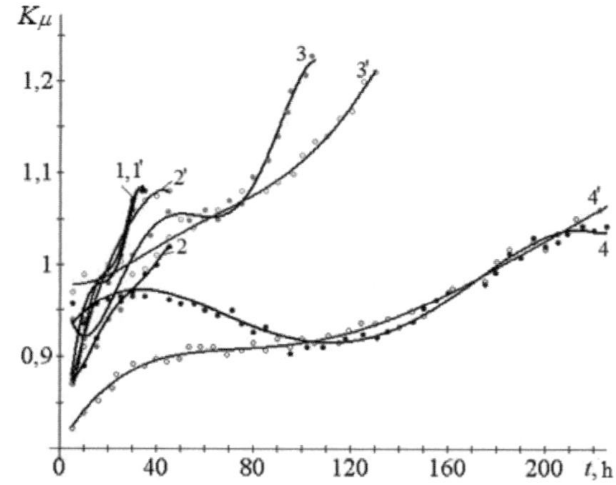

Figure 2. Dependence of the viscosity relative factor on time and temperatures of mineral motor oil testing M10-Г$_2$к (see figure 1).

As can be seen in the functional dependence, as the thermostatting time increases, viscosity grows for each of temperatures. At high temperatures (curves 1, $1^/$–2, $2^/$) an intensive increase in viscosity is observed that specifies high speed of formation of insoluble oxidation products. On the other hand, high temperature models aggressive operation conditions of the machine or an assembly, and consequently, high temperature of oil in micro volume.

A similar tendency of the increase in viscosity is shown at long testing time at temperatures of 160°C and 150 °C (curves 3, $3^/$ and 4, $4^/$), that is typical of mineral motor oils and reflects destructions of viscous additives which are found in the basis of the mineral oil.

Dependences of volatility on time and temperatures of the test (Fig. 3) characterize the presence of distillate fractions in oil.

High volatility molecules of a lubricant is observed at temperatures 180°C and 170°C, which is caused by high speed oxidizing processes and evaporation of easy fractions. At the subsequent temperatures more flat dependences (curves 3, $3^/$ and 4, $4^/$) describe gradual process of distillate fractions evaporation of lubricant oil.

Influence of metals on thermal-oxidation stability of lubricants is determined by the factor of the catalytic actions K_c in the expression:

$$K_C = S_{Ka} / S_{Ka.c}, \quad (3)$$

where S_{Ka} is the area limited by the curve of the dependence of the factor of the light flux absorption K_a on time in testing the commodity lubricant without catalyst; $S_{Ka.c}$ is the area limited by the curve of the dependence of the factor of the light flux absorption K_a on time in testing the commodity lubricant, with catalyst.

A quantity indicator of the catalytic action of steel on the oxidizing processes proceeding in the lubricant is determined under the Equation (3). It is established, that at the value $K_c > 1$, the process of motor oil oxidation M-10-G$_{2к}$ is accelerated, and at the value $K_c < 1$ slows down.

Hence, at the accelerated process of oxidation, as a result of the shipped metal in the tested oil, steel 45 is the catalyst of oxidation, that is the additional factor promoting acceleration of the ageing processes of a lubricant at the value of factor K_c<1 - inhibitor oxidations.

4. Results of the testing the temperature stability of oils

Influence of temperature on oil optical properties change was estimated by the factor of the light flux absorption K_a (figure 4). The dependences of the light flux absorption factor on temperature have two characteristic sections of various intensity for mineral (curve 1), partially synthetic (curve 2) and pure synthetic (curve 3) oils are established. It means that the oils of different bases have distinctions in structure and concentration of the destruction product and their influence on optical properties. The second section is characterized by stabilization of factor K_a confirming the end of the destruction process.

The destruction of the mineral partially synthetic and pure synthetic oil occurs in the first section and can be described by the equations of the second order

$$K_a = \alpha_1' T^2 + \alpha_2' T + b', \tag{4}$$

where α'_1 and α'_2 are the factors describing the intensity of formation of destruction products; b' is the factor dependent on the basis.

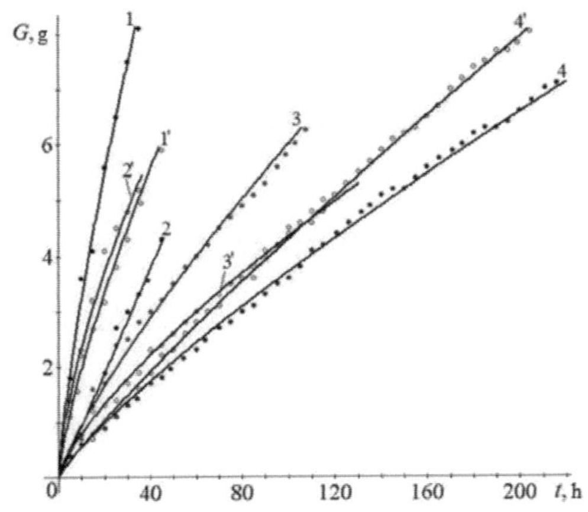

Figure 3. Dependence of volatility from time and temperature of test (see figure 1).

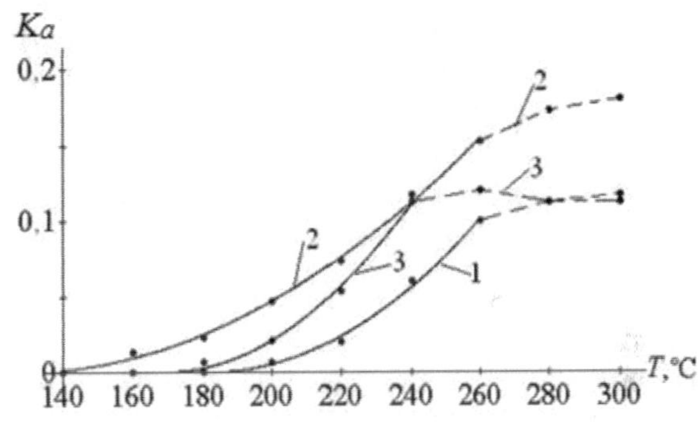

Figure 4. Dependence of the factor of light absorption flux K_a on temperature T in testing oils: mineral Mobil 10W-40 SJ/CH (1); in part synthetic Zic 5000 10W-40 CG-4/SH (2); synthetic Pentosynth 5W-40 SH/CF (3).

The factors of the destruction process are the values of temperatures at the beginning of destruction and its end, so, for mineral oil (curve 1) they are 185 °C and 260 °C, respectively, for partially synthetic oil (curve 2), they are 140 °C and 260 °C, and for pure synthetic oil (curve 3), they are 170°C and 240 °C.

The speed of the destruction as a quantity indicator of the destruction process for oils of various bases is determined and described by the derivatives of the first order Equations (5)

$$V_{Ka} = 2 \cdot \alpha T + c \tag{5}$$

where α is the factor describing a corner inclination of dependence $V_{Ka}=f(T)$; c is the factor describing destruction resistibility of oils; T is temperature of test, [°C].

With increase in temperature of test from 200 °C, reduction of viscosity that testifies about destruction a basis of oils is observed.

We suggest, that in destruction process with the increase in temperature optical properties of oil, speed of destruction process change, therefore, the offered complex criterion of temperature stability can be determined by the expression

$$K = V_{Ka} \cdot (T_2 - T_1), \quad (6)$$

where V_{Ka} is the destruction speed of the oil, 1/°C; T_1 is the temperature when destruction started, °C; T_2 8 is limiting temperature of the destruction, °C.

The more the temperature stability of the lubricant, the less is the value of factor *K*. The value of factor *K* is used to provide the operational properties:

at $K<0,15$, oils belong to group SL;

$0,15 \leq K<0,3$, oils belong to group SJ;

$0,3 \leq K<0,5$, oils belong to group SH;

$0,5 \leq K<0,8$, oils belong to group SG;

$K\geq 0,8$, oils belong to group SF.

The given criterion allows identification of oils by the groups of operational properties.

5. Conclusion

Methods for determining the temperature stability and thermal-oxidation stability of lubricants allow us to determine a new factor for estimation of destruction processes and oxidation of commodity lubricants which extends information on their quality, helps to make the right choice of the design stage of machines and units and to improve the system of classification and identification of oils by the groups of operational properties.

An integrated criterion of the catalytic action of metals on oxidizing processes of the lubricant is offered. It has been identified that at high temperatures, chemical adsorption of surface-active substances of the additives alloying the base oil occurs, and the processes of blanket formation proceed intensively. Thus, the superficial energy of the solid material influence the processes of self-organizing tribology systems, and the lubricant, one of the constructive elements of units and machines, is currently used.

References

[1] ASTM D5704-15a 2015 *Standard Test Method for Evaluation of the Thermal and Oxidative Stability of Lubricating Oils Used for Manual Transmissions and Final Drive Axles*

[2] ASTM D5763-11 2011 *Standard Test Method for Oxidation and Thermal Stability Characteristics of Gear Oils Using Universal Glassware*

[3] ASTM D6557-13 2013 *Standard Test Method for Evaluation of Rust Preventive Characteristics of Automotive Engine Oils*

[4] ASTM D2266-01 2008 *Standard Test Method for Wear Preventive Characteristics of Lubricating Grease (Four-Ball Method)*

[5] Bezborodov Yu N *et al* 2011 *Methods of control and diagnostics of the performance properties of lubricants in the parameters of thermo-oxidative stability and temperature resistance* (Krasnoyarsk: Siberian Federal University)

IOP Conf. Series: Materials Science and Engineering **132** (2016) 012027 doi:10.1088/1757-899X/132/1/012027

The Classification of Weld Seam Defects for Quantitative Analysis by means of Ultrasonic Testing

Y Salchak[1], T Tverdokhlebova[2], S Sharavina[3] and A Lider[4]

[1]Junior Researcher, National Research Tomsk Polytechnic University, Tomsk, Russia
[2]Student, National Research Tomsk Polytechnic University, Tomsk, Russia
[3]Engineer, National Research Tomsk Polytechnic University, Tomsk, Russia
[4]Head of Department, National Research Tomsk Polytechnic University, Tomsk, Russia

E-mail: yasalchak@gmail.com

Abstract. The paper describes effective quality assessment of spent nuclear fuel storage cask. The ultrasonic testing method is considered. The classification of possible defects with corresponding dimensions limits is proposed. The database of defects of the spent nuclear fuel storage cask was created in compliance with the nuclear energy industry regulatory documents.

1. Introduction

Nowadays, nuclear energy is widely used all over the world due to the low level of fuel consumption, low price, and high capacity. It is the most promising way to produce energy [1–2].

However, all stages of the nuclear fuel cycle give rise to some nuclear waste that also requires specific storage conditions. There are two types of spent nuclear fuel (SNF) storage: dry and wet storage. Dry SNF storage is a current state-of-the-art technology. Mining and Chemical Enterprise (MCE) is one of the leading Russian companies that deals exclusively with the dry storage of SNF [3].

MCE has developed a special pressure-tight cask for the dry storage of SNF. It is a sealed container made of austenitic steel. In order to provide safety, an effective casks quality assurance method is required. At present, X-ray testing is used for this purpose. However, X-ray is a time consuming and risk bearing procedure due to potential radiation exposure. Nondestructive acoustic techniques can be applied as an effective nonradioactive method of monitoring. For example, ultrasonic method can provide proper evaluation of the degradation process of structural material of the cask and its weld seam [4]. Although ultrasonic testing is the relative method of measurement, it requires calibration block with relevant artificial defects to adjust the equipment [5]. It is important to analyze the structure and identify the parameters of an object of control before developing calibration block. Possible defects of the control object are to be determined. Therefore, the goal of this work was to create the database which contains SNF cask defects classification.

2. Object of control and defects classification

The cask for SNF dry storage was chosen as an object of control. It is the novel technology of nuclear power industry developed by MCE. This cask is intended for the storage of spent nuclear fuel of uranium-graphite channel type reactor RBMK [6].

Table 1 represents basic characteristics of the object of control.

Table 1. Characteristics of the SNF storage cask.

Structural material	Dimensions [mm]	Types of seal welds	Welding
austenitic steel X10CrNiTi18-10	length – 1000 diameter – 635 thickness- 4	girth longitudinal	automatic argon-arc welding

Reliable quality assessment is achieved during weld seam testing. Welds are chosen as an inspection area due to flow growth which is most probable in this area. The cask consists of three main parts which are hermetically sealed. There are two types of seal welds: girth and longitudinal welds (figure 1).

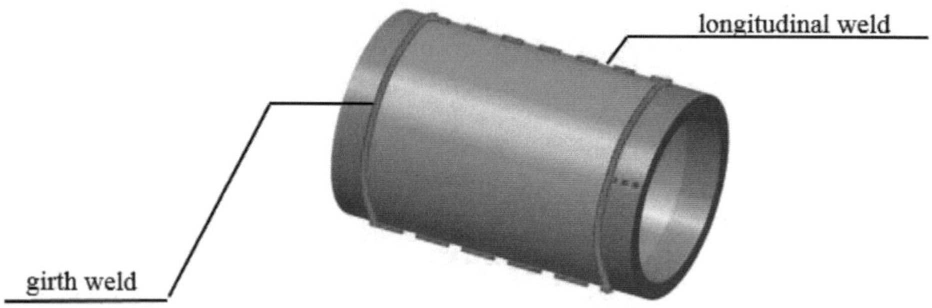

Figure 1. SNF cask`s body.

Quality control is applied to monitor the quality of the casks produced. The main purpose is flaw detection, determination of its size and orientation. There are two kinds of defects in respect to the time of formation: defects that occur during manufacturing process and those that occur during exploitation of the produced item.

Defects generated due to production are caused by breakdowns in the technological process. For welds, this group of defects comprises metal inclusions, blowholes and tungstic splashes. Another group of defects occurs due to the overload conditions (of exploitation nature only) or breakdowns in operating conditions.

This group of defects mostly comprises different types of cracks: fatigue cracks, stress cracks, overload-induced cracks, stress corrosion cracks and corrosion fatigue cracks.

For the considered object of control, there are several types of defects caused by the particular welding type of construction material. The cask is made of austenitic steel, X10CrNiTi18-10. This material has high resistance to embrittlement and corrosion, and good weldability. However, welding of X10CrNiTi18-10 steel may cause grain-boundary corrosion and transform into a defect during cask exploitation [7–8].

Furthermore, aqueous conditions at the surface, presence of aggressive species, heavy loads and changes of temperature cycles also give rise to material degradation processes that lead to penetration of a cask. As a result, there could be different failure modes such as porosity, cracks and others. The detectability is very important to enable the effective quality control for each failure mode and defect [9].

At the same time, all components of the nuclear power industry are made in compliance with the industrial standards. In accordance with these documents, the quality requirements for each type of the item depend on extent of the consequences in case of its failure. Usually, these requirements depend on the purposes of the item and thus specify frequency of inspections. For example, such parts as turbine blades, valve in the cooling system of NPP and SNF storage casks belong to the most critical groups according to the quality scale [10]. It is necessary to provide the effective quality assurance of these components to ensure that parts meet the requirements of specifications.

Quality control of mechanical structure of such components could be realized by means of nondestructive testing [11]. Ultrasound method of control was chosen for the present work. The main goal is to estimate if a component is still reliable in case of some discontinuity flaws.

Due to the technical regulatory documents [12], there could be different types of welding defects. The main parameters to determine the component reliability are quantity characteristics of discontinuity in accordance with the component quality group. Thus, for each component there should be a classification of possible defects that provides information about permissible imperfection size.

In the present work all typical defects of the SNF storage casks are classified. The classification comprises pores, non-metallic inclusions, lack of fusion, incomplete penetration, variations in weld geometry and others. According to this classification the database of possible defects was developed.

3. The database of the SNF storage cask`s defects

The database was implemented using FileMaker Pro Advanced software. It is represented in the table that comprises all possible variables (figure 2). The operator can add new data to extend the database. This function is necessary if there is a new object of control which means the operator should consider other types of defects.

The table demonstrates all symbols used to identify type, parameters (def_type – type of the defect; def_kind – category of a defect; ref – reference sample), permissible dimensions (*h* – height, *d* – diameter; *I* – length; *L* – distance between two defects), schematic and realistic images of imperfections (figure 2).

The database can be used both for automatic and manual ultrasonic testing. This system comprises complete information about all possible defects and imperfections including their permissible dimensions depending on the required quality level of the weld.

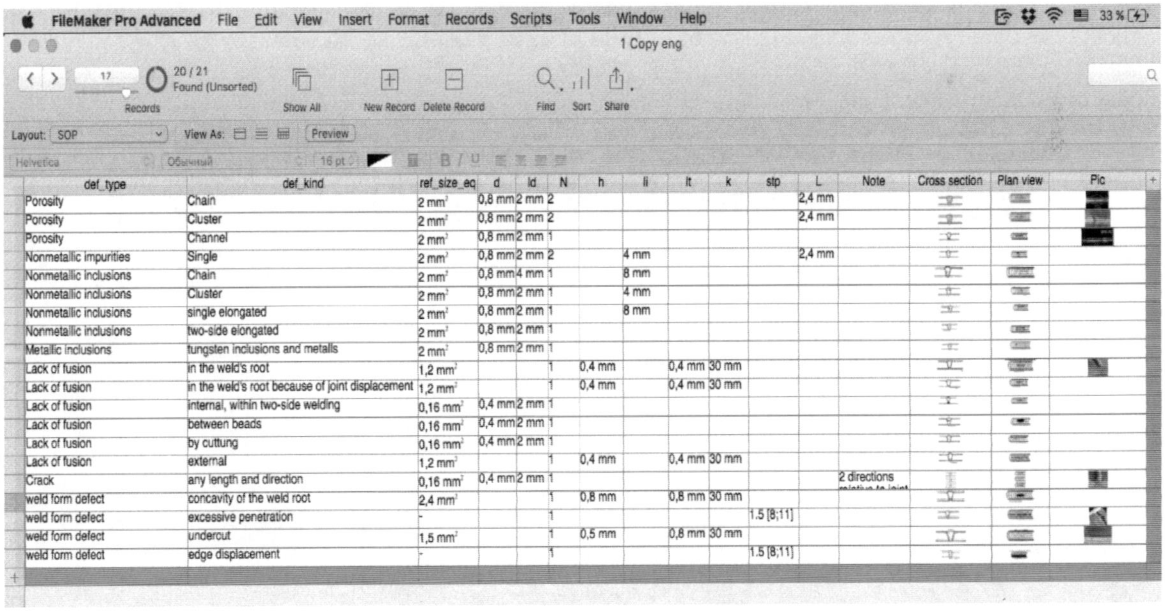

def_type	def_kind	ref_size_eq	d	ld	N	h	li	lt	k	stp	L	Note	Cross section	Plan view	Pic
Porosity	Chain	2 mm²	0,8 mm	2 mm	2						2,4 mm				
Porosity	Cluster	2 mm²	0,8 mm	2 mm	2						2,4 mm				
Porosity	Channel	2 mm²	0,8 mm	2 mm	1										
Nonmetallic impurities	Single	2 mm²	0,8 mm	2 mm	2		4 mm				2,4 mm				
Nonmetallic inclusions	Chain	2 mm²	0,8 mm	4 mm	1		8 mm								
Nonmetallic inclusions	Cluster	2 mm²	0,8 mm	2 mm	1		4 mm								
Nonmetallic inclusions	single elongated	2 mm²	0,8 mm	2 mm	1		8 mm								
Nonmetallic inclusions	two-side elongated	2 mm²	0,8 mm	2 mm	1										
Metallic inclusions	tungsten inclusions and metalls	2 mm²	0,8 mm	2 mm	1										
Lack of fusion	in the weld's root	1,2 mm²			1	0,4 mm		0,4 mm	30 mm						
Lack of fusion	in the weld's root because of joint displacement	1,2 mm²			1	0,4 mm		0,4 mm	30 mm						
Lack of fusion	internal, within two-side welding	0,16 mm²	0,4 mm	2 mm	1										
Lack of fusion	between beads	0,16 mm²	0,4 mm	2 mm	1										
Lack of fusion	by cuttung	0,16 mm²	0,4 mm	2 mm	1										
Lack of fusion	external	1,2 mm²			1	0,4 mm		0,4 mm	30 mm						
Crack	any length and direction	0,16 mm²	0,4 mm	2 mm	1							2 directions			
weld form defect	concavity of the weld root	2,4 mm²			1	0,8 mm		0,8 mm	30 mm						
weld form defect	excessive penetration	-			1					1.5 [8;11]					
weld form defect	undercut	1,5 mm²			1	0,5 mm		0,8 mm	30 mm						
weld form defect	edge displacement	-			1					1.5 [8;11]					

Figure 2. Table of variables.

All the values used in the table were chosen in accordance with the regulatory government standards. The permissible dimensions depend on the type of defect chosen by the operator (figure 3).

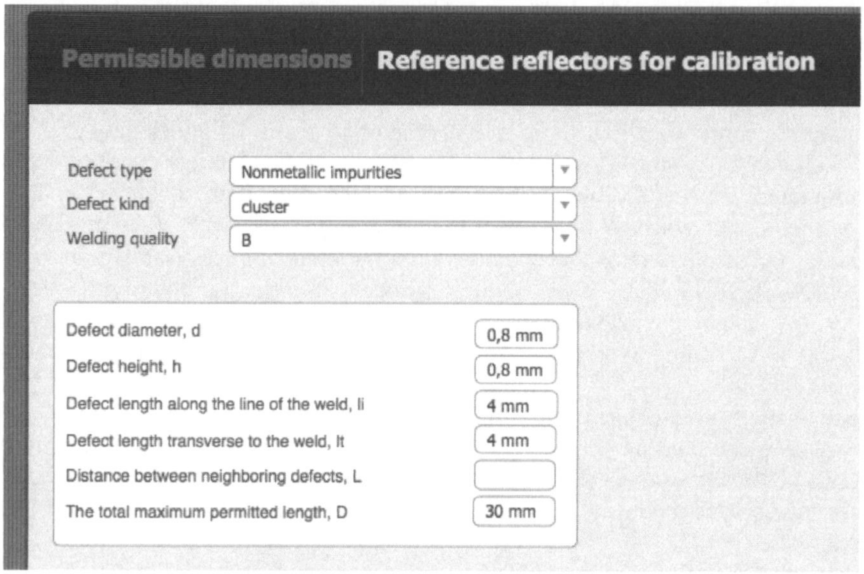

Figure 3. Operator`s report window.

There is a particular reference reflector for each type or category of defects. The dimensions of these reflectors should correspond to the parameters specified in the regulatory documents. Thus, using the created database the operator can estimate whether the detected imperfection is permissible for a given level of weld quality. Moreover, the operator can select appropriate reference reflector for the SNF storage cask. Reference reflectors are used for calibration and setting the ultrasonic testing system. The operator should use the type and category of a defect in order to get necessary information about reference reflectors for designing of a calibration block (Fig. 4). The developed database makes this process highly effective. Moreover, it provides images that demonstrate location of each particular type or category of defects.

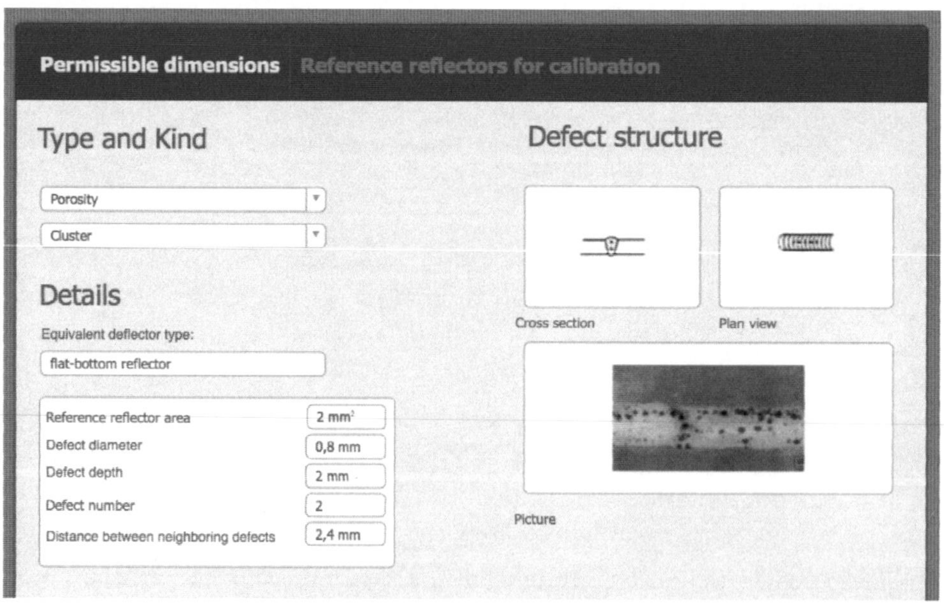

Figure 4. Characterization of calibration block.

4. Conclusion

The developed database provides all necessary information about defects that occur in the SNF storage cask. It includes information about defect type, its category and permissible dimensions in compliance with the technical regulatory documents.

In order to provide an effective quantitative analysis, a calibration block should be designed. It should contain reference reflectors inside. These reflectors are to relate to the real defects of the SNF storage cask. The defect database allows selecting appropriate parameters of reference reflectors for each level of weld quality.

Due to implementation of quantity estimation ability, the results of the present work provide the same level of measurement accuracy of ultrasonic testing as the level of X-Ray testing. Furthermore, ultrasonic testing is much safer and less time-consuming method of control. These facts emphasize the advantages of ultrasonic testing and within advanced techniques of ultrasonic industrial tomography enable to significantly increase the capacity of SNF cask manufacturing.

Within ongoing research in this project, it is planned to extend the developed database. The images of control of real SNF cask with indications corresponding to particular defects will ensure higher reliability of ultrasonic testing. It will provide simpler and more accurate identification of detected flaws.

Acknowledgements

Research has been conducted with the financial support of Federal Target Programme, the project unique identifier is RFMEFI57514X0048.

References

[1] Dones R *et al* 2004 *Encyclopaedia of Energy* **3** 77-95
[2] IAEA 2013 *Climate change And Nuclear Power 2013* (Vienna: IAEA)
[3] Salchak Y *et al* 2015 *Physics Procedia* **70** 484–487
[4] IAEA Training Guidelines in Non-destructive 2012 *Testing Techniques Leak Testing at Level 2* (Vienna: IAEA)
[5] PNAE G-7-008-89 1989 *Regulations for design and safe operation of equipment and piping of atomic power plants* (Moscow)
[6] Gavrilov P *et al* 2012 Patent RU 2500 045 C1 *Byull. Izobret.* **33**
[7] Lancaster J. F 1984 *Physics in Technology* **15** 73 – 79
[8] Iida K 1996 *Nuclear engineering and design* **166** 85–98
[9] Technical Report 2013 *Failure Modes And Effects Analysis (FMEA) Of Welded Stainless Steel Canisters For Dry Cask Storage Systems* (Palo Alto: Electric Power Research Institute)
[10] PNAE G-01-011-97 1997 *General provisions for security of atomic power stations* (Moscow)
[11] Bolotina I *et al* 2013 *Russian Journal of Nondestructive Testing* **49** **(3)** 145–158 DOI: 10.1134/S1061830913030030
[12] GOST R ISO 6520-1-2012 1997 *Welding and Allied Process. Classification of Geometric Imperfections in Metallic Materials. Part 1. Fusion Welding* (Moscow: Standartinform)

IOP Conf. Series: Materials Science and Engineering **132** (2016) 012028 doi:10.1088/1757-899X/132/1/012028

Modelling of Fuzzy Control Modes for the Automated Pumping Station of the Oil and Gas Transportation System

A Sagdatullin[1], E Muravyova[2] and M Sharipov[3]

[1]Engineer, Research and Educational Center, Almetyevsk, Russia
[2]Professor, Research and Educational Center, Almetyevsk, Russia
[3]Lecturer, Research and Educational Center, Almetyevsk, Russia

E-mail: saturn-s5@mail.ru

Abstract. An attempt has been made to explain to the scientific and technical establishment the fact that the fuzzy controllers with input/output parameters represented by the set of precise terms of automate control objects, which operative algorithm is verbally presented as a unique experience of experts in the domain, in simpler way, with higher speed, slighter error and less costs as compared with the standard fuzzy controllers.

1. Introduction

Pumps and pumping plants of collection systems, seam pressure maintenance, transportation and preconditioning of oil are important for the oil and gas sector as they provide:

- - collection and primary preconditioning of wells at the booster pumping plant (BPP), in close proximity to the automated group measuring units (AGMU);
- - preconditioning of process water on the initial water separation installation (IWSI) and its further pumping to cluster pumping plants (CPP);
- - metrological indicators of measuring units of product quantity and quality of wells at each stage;
- - transportation of commercial oil from the oil preconditioning unit (OPU) into the main pipelines by means of oil-pumping plants (OPP) located at distances up to 15 km.
- Consequently, the problem of energy saving and higher efficiency of pumping plants of collection system, seam pressure maintenance, transportation and preconditioning is topical. The solution of the problem is possible through the development and application of new automation systems for the above mentioned processes, because the analysis of traditional schemes of process control at the refining and petrochemical enterprises based on proportional-integral-differential (PID) controllers showed that the traditional PID controllers are not sufficiently effective for most of the complicated control objects owing to a number of features of the process progress [1–8].

2. Conceptual model of booster pumping station

The process diagram of the said pumping plant incorporates two asynchronous electric motors (AM), frequency converter, microprocessor controller, centrifugal pumps (main P–1 and standby P–2), pipe fittings and tank (E–3).

The programmable logic controller is the basis of BPP automation system and it performs both mathematical and logical operations which are necessary to either control the oil transport process or to perform control (PID controller) depending on the progress dynamics of the plant operation processes. However, PID controllers, which are the basis of automated electric drive systems with frequency converters, cannot provide high quality control by nonlinear technological processes which include BPP objects. Consequently, energy consumption increases and energy-saving effect is greatly reduced due to the introduction of VFD. Figure 1 shows a conceptual model of booster pumping plant.

Input variables are Q_6 is flow rate through valve 6 (figure 1) of the incoming emulsion to the tank from BPP-1s, Q_{17} is flow rate through valve 17 of the incoming emulsion to the tank from POGP-1, Q_{18} is flow rate through valve 18 of the incoming emulsion to the tank from tank VST-1, ω_{out1} is angular velocity at the outlet of the main pump, ω_{out2} is angular velocity of the output standby pump, Q_Σ is total flow rate of the oil emulsion incoming to the BPP.

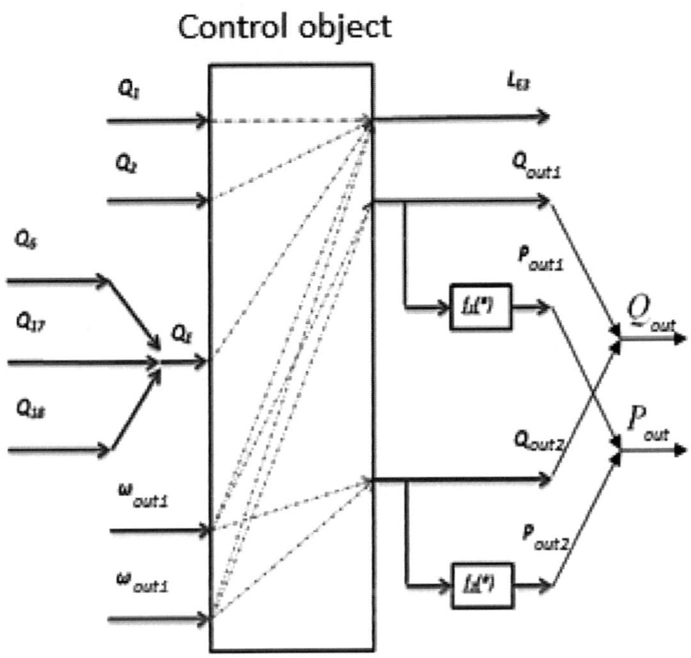

Figure 1. Conceptual model of a booster pumping station, where: Q_{out}, P_{out} – flow rate and pressure at the BPP outlet made up of individual flow rates Q_1, Q_2 and pressure values P_1, P_2 – on pumps *P*-1 and *P*-2 outlets.

BPP output variables are: the liquid level in the tank L_{E3}; Q_{out1}, P_{out1} is the flow rate and pressure respectively on the outlet of the main pump *P-1*; Q_{out2}, P_{out2} are the flow rate and pressure respectively on the outlet of the standby pump *P-2*; $f_1(*)$ and $f_2(*)$ are the conversion function P_{out1} and P_{out2} depending on Q_{out1} and Q_{out2}, Q_1, Q_2 are feedback channels, composed of separate flow rate values Q_{out1} and Q_{out2} on pump *P*–1 and *P*–2 outlets.

The dependence of the pump pressure head and flow rate on the angular velocity of the motor is described by the following nonlinear expression:

$$H = \frac{H_F}{\omega_N^2}\omega^2 - S_F Q^2, \qquad (1)$$

where H_F is hypothetical static pressure head, m; S_F is hypothetical hydraulic pump resistance, s^2/m^5; Q is pump feed (amount of fluid pumped by the pump per unit of time), m^3/s; ω and ω_N are current and nominal rotational speed of the pump shaft, respectively. Figure 1 shows that output function $L_{E3} = f\,(Q_6, Q_{17}, Q_{18}, Q_1, Q_2, \omega_{out1}, \omega_{out2})$ is the function of seven arguments whereas functions P_{out1} =

$f_1(\omega_{out1}, \omega_{out2})$, $P_{out2} = f_2(\omega_{out1}, \omega_{out2})$, $Q_{out1} = f_3(\omega_{out1}, \omega_{out2})$ and $Q_{out2} = f_4(\omega_{out1}, \omega_{out2})$ are those of two arguments. Therefore, the pumping plant in question can be characterized as a multilinked nonlinear control object with the dimension equal to three.

It follows that the booster pumping plant controlled parameters are: level in the tank L_{E3}, pressure (P_{out1}, P_{out2}) and flow rate (Q_{out1}, Q_{out2}) in the output pipelines. The level in the BPP tank should be maintained at 2.5 meters (tank height is 5 m, length is 10.2 m, and capacity is 200 m^3).

3. Logical diagram of pumping station process control

The prerequisites to improve the control quality of pumping plant are the fast response of the electric motor speed and controlled valves displacement to external interferences and assurance of certain preset values of the controlled parameters (flow rate, pressure head and level).

The continuous variation of the oil supply to the pumping plant leads to oscillations in oil pressure and flow rate in the pipeline, which reduces the control quality. Consequently, PID controllers require periodic and time-taking adjustment causing the increase in operating costs of the control system. To boost the energy efficiency and control quality, it is proposed to use a multi-dimensional discrete logic controller, which input and output variables are represented by the set of terms of rectangular shape of the membership function, i.e. discrete terms [9–12].

Electric motors and valve K-33 are controlled by the signals from microprocessor controller (MPC), which receives information from the sensors: motor angular speed, position of the valve K-33 actuator (valve closing position as percentage), pressure and liquid level in the tank L_{E3}. The level in the PP tank should be maintained at 2.5 meters (tank height is 5 m, length is 10.2 m, and capacity is 200 m^3). Logical diagram of the process control algorithm is shown in figure 2. As a result, controllers w_1, w_2, and K-33 are come in action when the transition condition are true (L_{E3}> 2.5 m) and (L_{E3} >> 2.5 m).

As can be seen from figure 3, in such a program for each control loop, along with controlling ($PRContS_{Q20}$, $PRContS_{Q21}$, $PRContS_{Q\text{-}33}$), there are also compensation ($PRCS_{Q20}$, $PRCS_{Q21}$, $PRCS_{Q\text{-}33}$) production rule systems. Thereby, the interference of multidimensional discrete logic controller loops is minimized and the control quality of liquid level in the tank is increased.

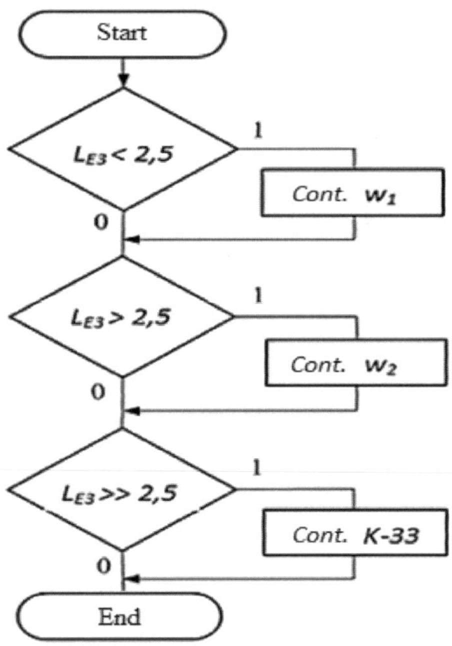

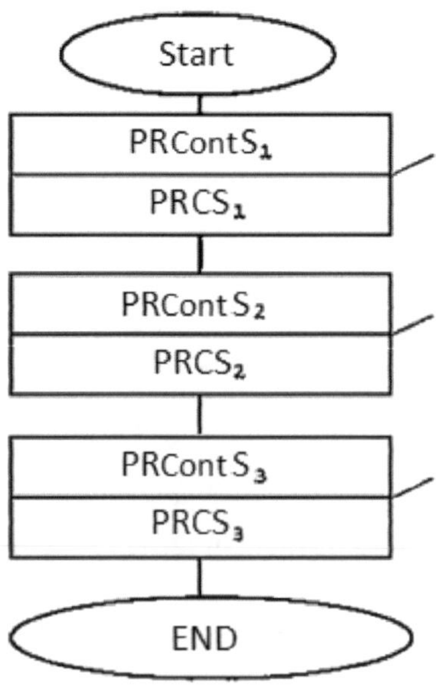

Figure 2. Logical diagram of pump station control process.

Figure 3. Logical structure of the software-based 3-dimensional discrete logic controller of the pumping station.

Figure 4 shows the interpretation of the controlled parameter L_{E3} by the set of 16 discrete terms $L_{E31}, L_{E32}, \ldots, L_{E315}, L_{E316}$ (e.g., $T_{LE31} < T_{LE3} < T_{LE316}$ etc.). Interpretation of the other adjustable controlled parameters Q_{out} and P_{out} is similar.

The analytical expression for the term set for L_{E3} variable is as follows:

$$T(L_{E3}) = \sum_{i=1}^{16} L_{E3} \cdot ((i-1) \cdot 0{,}25 \le L_{E3} < i \cdot 0{,}25 . \tag{2}$$

where i is the number of discrete term.

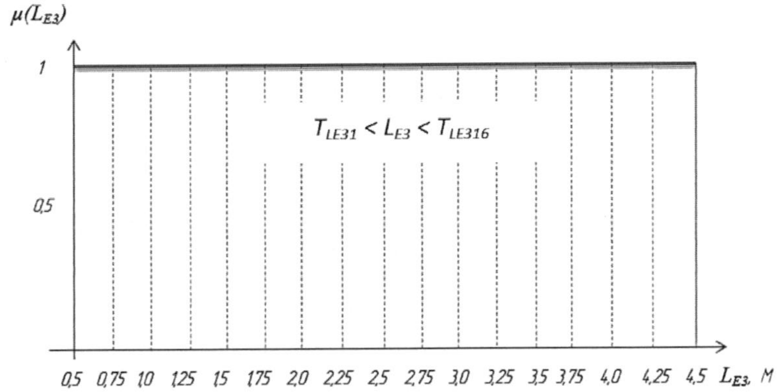

Figure 4. Location of discrete terms of the oil emulsion level in the tank along the universal numerical axis.

4. Interpretation of the compensation function by set of discrete terms

Figure 5 shows compensation function TLE3к, which analytical expression of the universal term sets corresponds to:

$$T = \sum_{i=1}^{16} T_{LE3\kappa i} \cdot ((i-1) \cdot 6 \le t_{TLE3\kappa} < i \cdot 6 \tag{3}$$

Introduction of compensation functions changed the structure of the program of pumping plant MPLC.

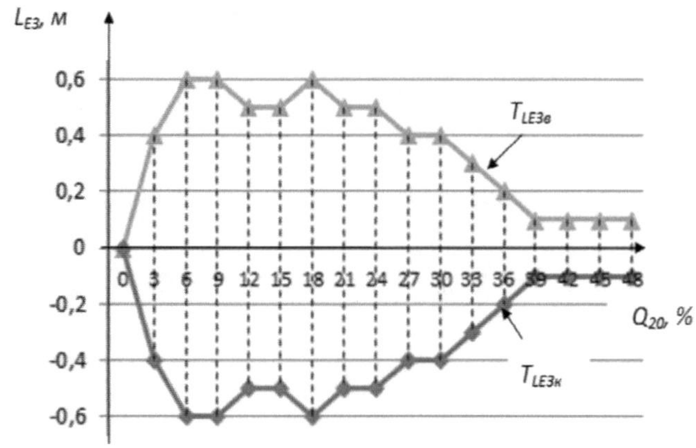

Figure 5. Interpretation of the compensation function $T_{LE3\kappa}$ by the set of discrete terms $T_{LE3\kappa 1} \div T_{LE3\kappa 16}$.

Replacement of the control (ω_1, ω_1, Q_{33}) and controlled (Q_{20}, Q_{21}, L_{E3}, Q_{33}, H_{out}) parameters of the pumping plant by the set of discrete terms allows us to construct a production rule system for the three-loop discrete fuzzy controller. For a loop with the controlled parameter $L_{E3,}$ the system has the following structure:

$$\begin{cases} If\ L_{E3} = L_{E312} v L_{E313} v ... v L_{E315} v L_{E316},\ then\ Q_{20} = 0; \\ If\ L_{E3} = L_{E311},\ then\quad Q_{20} = T_{Q201} v T_{Q20k1}; \\ \dots \\ \dots \\ \dots \\ If\ L_{E3} = L_{E32},\ then\quad Q_{20} = T_{Q2015} v T_{Q20k15}; \\ If\ L_{E3} = L_{E311},\ then\quad Q_{20} = T_{Q2016} v T_{Q20k16}. \end{cases} \quad (4)$$

According to this expression, the parameter value L_{E3} is maintained at the level of 2.5 m with the error of ± 0.25 m, and certain discrete terms $L_{E3\kappa 1} \div L_{E3\kappa 16}$ are introduced into production rule consequents to compensate the other control loops for parameter L_{E3}. The production rule system relating to controlled parameters L_{E3}, Q_{out}, P_{out} is built using the similar structure.

5. Conclusion

In this paper, the electric pumping station control system has been proposed based on the three-dimensional fuzzy logic controller, which input and output variables are represented by the set of terms of rectangular shape of the membership function. This system enables us to stabilize the oil level at 2.5 m with the absolute error of ± 0.2 m, as well as to improve the preconditioning quality of oil emulsion at the oil preconditioning plant.

References

[1] Zadeh L A 1965 *Information and Control* **8** 338–353
[2] Mamdani E H *and* Assilian S A 1975 *Int. J. of Man-Machine Studies* **7 (1)** 1–13
[3] Mamdani E H 1977 *IEEE Transactions on Computers* **26 (12)** 1182–1191
[4] Demenkov N P 2005 *Fuzzy control in engineering systems* (Moscow: Publishing House: Bauman Moscow State Technical University)
[5] Lee C C 1990 *IEEE Transactions on Systems, Man and Cybernetics* **20 (2)** 404–435

[6] Sagdatullin A M *et al* 2015 *Oil Industry* **3** 92–95
[7] Kaufmann A *and* Gupta M M 1985 *Introduction to fuzzy arithmetic: Theory and applications* (New York: Van Nostrand Reinhold)
[8] Yager R *and* Filev D 1994 *J. of Intelligent & Fuzzy Systems* **2(3)** 209–219
[9] Sagdatullin A M *et al* 2014 R.F. Patent 140350
[10] Sugeno M 1985 *Industrial Applications of Fuzzy Control* (Elsevier Science Pub. Co.)
[11] Sagdatullin A *et al* 2015 *Applied Mechanics and Materials* **756** 633–639 DOI:10.4028/www.scientific.net/AMM.756.633
[12] Zimmermann H J 1987 *Fuzzy Sets, Decision Making and Expert Systems* (Boston, Kluwer)

IOP Conf. Series: Materials Science and Engineering **132** (2016) 012029 doi:10.1088/1757-899X/132/1/012029

Calibration Interval Adjustment of a Measuring Instrument in Industries During Long-term Use

N Natalinova[1], N Ilina[2] and E Frantcuzskaia[3]

[1]Associate Professor, National Research Tomsk Polytechnic University, Tomsk, Russia
[2]Student, National Research Tomsk Polytechnic University, Tomsk, Russia
[3]Senior lecturer, National Research Tomsk Polytechnic University, Tomsk, Russia

E-mail: natalinova@tpu.ru

Abstract. Calibration interval adjustment of measuring instruments is one of the urgent tasks in industries. The article represents the calibration interval calculation of the potentiometer PCB-4P according to the verifications for the 4 year period of the Metrological Department in the aviation plant. The calibration interval is shown to be increased according to the calculation of its reliability and stability of metrological characteristics.

1. Introduction

A form of providing and maintaining the appropriate level of metrological reliability of measurements in industries is the calibration. To avoid the problems that inaccurate test equipment can cause, companies must calibrate their equipment regularly. Some companies, however, are so strict about sticking to their calibration schedules that they often calibrate equipment in excess. Although this helps them avoid producing bad parts, unnecessary calibrations also increase costs. The calibration is carried out periodically, at certain time intervals, which are called calibration intervals, established in the certification of measuring instrument evaluation according to the existing system of ensuring uniformity of measurements. Practically, during long-term use in industries it is necessary to make adjustments of the calibration interval of measuring devices. In some types of measuring devices one or even more of the metrological characteristics can go beyond the regulated value, established in the technical documentation, and thus bring the instrument to the disabled state. On the other hand, there are some measuring instruments which rarely demonstrate such kind of failures as metrological property instability of testing equipment, metrological failures and functional failures, so regular verification at certain intervals results in additional financial expense.

The procedure for determining and adjusting the calibration interval of measuring devices is a laborious and complicated task; consequently, nowadays there is hardly a solution for calibration interval adjustments in industries to be easily found.

The purpose of this work is to adjust the period of time between calibration sequences of the potentiometer type PCB-4P during long-term use in industries [1–8].

2. Methods

The process of adjustment of calibration intervals is quite time consuming though several stages can be distinguished such as:

1. Forming the measurand of measuring instruments for testing. The measurand quantity to be subjected to the test should be a minimum of 30. The same sample may include measuring instruments of various types but close in their purpose, construction, manufacturing technology or application conditions. After checking the uniformity of measuring instruments, the results of their test may be included into the same series of results.

2. The selected sample of measuring instruments is subjected to the test with either normal or accelerated mode (with the definite acceleration ratio). The measurements of the tested parameters are taken with equal periods of service or operation.

3. Considering the simple linear model of predicting, the method of least squares requires a minimum of 3 groups of multiple metering. Thus the test duration is to be not less than 2Δt, where Δt is the initial calibration interval.

4. Covering the results of instability which exceeds the intervals, the sample characteristic of instability distribution of metrological features of measuring instruments is found: the values of the sample mean and sample standard deviation.

5. The values acquired enable to evaluate the time-dependence function of the values of the sample mean and sample standard deviation. Polynomial equation coefficients are selected with the method of least squares. The polynomial order is selected from the options 1…5 according to the minimum error approximation criterion [9–10].

The initial calibration interval for a new piece is important, taking into account the manufacturer's recommendations. This is especially true if there has not been any prior experience with a particular piece of equipment. Though having had any similar equipment already in service, we can use our experience with that equipment to set the calibration interval for the new piece of equipment.

Initial calibration period recommended by the manufacturer for a new piece of equipment is to be taken in consideration in case if the equipment has not been operated before. However, the data gathered in the course of running the same piece of equipment in industries regarding its tests and measurements can also be used for adjusting the calibration interval for the new piece of equipment [11–18].

3. Results

The verification data of thirty devices PCB-4P for four-year period of running are listed in table 1.

According to manufacturer's recommendations, the calibration interval for this type of equipment is twelve months and the accuracy class is 0.5 %.

In order to correct the calibration interval, the results are grouped according to serial numbers of checks, since the manufacture or repair of the measuring devices:

- Group 1 – measuring instruments received in the first verification after manufacture or repair;
- Group 2 - measuring instruments received in the second verification after manufacture or repair;
- other groups according to the serial number of checks, manufacture or repair.

To carry out the statistical processing of the results of the coverage interval for each group enables to find options Fj…j= 1. 2 ... N. of the generalized normal distribution of metrological properties (instability of a measuring instrument), which the results of verification test equipment of this type match with the highest level of significance.

Table 1. Verification results of PCB-3P for 4 year period.

Number of Measuring Instrument	Values of metrological properties per year				Number of Measuring Instrument	Values of metrological properties per year			
	1	2	3	4		1	2	2	3
1	0.22	0.22	0.23	-0.2	16	0.35	0.11	0.15	0.16
2	0.22	0.21	-0.53	-0.13	17	0.37	-0.08	-0.06	0.11
3	0.23	0.2	-0.14	-0.2	18	0.11	0.1	-0.11	0.08
4	0.2	0.11	-0.17	0.53	19	0.11	0.06	0.14	0.09
5	-0.22	0.13	0.18	-0.13	20	0.26	0.12	0.51	0.12

6	0.22	-0.14	-0.15	-0.12	21	0.11	0.1	0.11	0.12
7	0.2	0.15	0.2	0.18	22	0.29	0.04	0.06	0.12
8	0.2	-0.2	-0.14	-0.2	23	0.35	-0.09	0.24	0.1
9	0.14	0.13	0.12	-0.15	24	0.12	0.19	-0.2	0.18
10	0.38	0.14	-0.09	-0.18	25	0.45	0.14	0.06	0.24
11	0.23	0.06	0.07	0.06	26	0.24	0.17	-0.11	-0.19
12	0.15	0.14	0.16	-0.07	27	0.21	-0.09	0.2	0.2
13	0.16	-0.08	0.11	0.16	28	0.3	0.22	-0.11	0.56
14	0.16	-0.12	0.1	-0.11	29	0.32	0.2	-0.13	0.18
15	0.26	-0.51	0.1	0.06	30	0.24	0.23	-0.11	0.1

To calculate this parameter the matrix of F values from 0 to 4 in 0.1 is defined.
For each value of the matrix we find:
Value X^F_i=sign ($x_i|x_i|$) = 1 ... N;

where sign(x_i), $\begin{cases} X_i < 0, \text{sign}(x_i) = -1 \\ X_i = 0, \text{sign}(x_i) = 0 \\ X_i > 0, \text{sign}(x_i) = 1 \end{cases}$

The calculation results are shown in table 2.

Table 2. The parameter values of the generalized normal distribution of PCB-3P metrological properties.

Number of parameter	F / x_i	0	0.1	0.2	0.3	…	2.9
1	2	3	4	5	6	7	8
1	0.22	1	0.8595	0.7387	0.6349	…	0.0124
2	0.22	1	0.8595	0.7387	0.6349	…	0.0124
3	0.23	1	0.8633	0.7453	0.6435	…	0.0141
4	0.2	1	0.8513	0.7248	0.6170	…	0.0094
5	-0.22	1	-0.8595	-0.7387	-0.6349	…	-0.0124
6	0.22	1	0.8595	0.7387	0.6349	…	0.0124
…							
29	0.32	1	0.8923	0.7962	0.7105	…	0.0367
30	0.24	1	0.8670	0.7517	0.6517	…	0.0159

According to Table 2, the values of the sample mean and sample standard deviation obtained due to of the equations (2) and (3). The results are shown in table 3.

Table 3. Values of the sample mean and sample standard deviation.

F	0	0.1	0.2	0.3	0.4	0.5	0.6	0.7
m	1	0.8022	0.6905	0.5953	0.5139	0.4443	0.3847	0.3336
σ	0	0.3156	0.2758	0.2435	0.2168	0.1945	0.1755	0.1590
F	0.8	0.9	1	1.1	1.2	1.3	1.4	1.5
m	0.2897	0.2519	0.2193	0.1912	0.1670	0.1459	0.1277	0.1119
σ	0.1446	0.1318	0.1203	0.1100	0.1007	0.0922	0.0845	0.0774
F	1.6	1.7	1.8	1.9	2	2.1	2.2	2.3
m	0.0982	0.0863	0.0759	0.0668	0.0589	0.0520	0.0459	0.0406
σ	0.0709	0.0650	0.0596	0.0546	0.0500	0.0459	0.0420	0.0385
F	2.4	2.5	2.6	2.7	2.8	2.9		
m	0.0359	0.0318	0.0282	0.0250	0.0223	0.0198		
σ	0.0353	0.0324	0.0297	0.0272	0.0249	0.0229		

The value of the likelihood function is calculated according to the equation:

$$\ln[L(F)] = N\ln\left[\frac{|F|}{\sqrt{2\pi\sigma_i}}\right] + (F-1)\sum_{i=1}^{N}\ln(|x_{il}|) - 0{,}5(N-1), \quad (1)$$

where F – distribution exponent.

To find the maximum of likelihood function such steps are to be made. As F_j the value corresponding to ln[L(F)] the equation (1) is taken. According to the principle of maximum likelihood function, it will be the best approximation of the sampling distribution of metrological properties (instability of a measuring instrument) due to the normal distribution law.

The maximum value of the function ln [L (F)] in the first group reaches F = 1.1.

Next, the distribution of values of optional features X^F_i (jT) for every group of measuring potentiometer PCB-4P are to be found:

– the value of the sample mean is accounted according to the equation:

$$\overline{m}_i(T_j) = \frac{1}{N}\sum_{i=1}^{N} X_i^F(T_j),\ j = 1{,}2...; \quad (2)$$

where T_J - the order of calibration interval;

– the value of the sample standard deviation is calculated according to the equation:

$$\sigma_i(T_j) = \sqrt{\frac{\sum_{i=1}^{N}(x_i^F(T_j) - m_i(T_j))^2}{N-1}},\ j = 1,\ 2...N \quad (3)$$

The results of data processing for the four PCB-4P Verifications are shown in Table 4.

Table 4. Sample characteristics of the values distribution of X^F_i (jT).

	Sample mean	Sample standard deviation	Maximum of ln [L(F)]	F
Group 1	0.0084	0.1122	18,13	1.1
Group 2	0.0404	0.1256	13,53	1.2
Group 3	0.0166	0.1536	7,7	1.2
Group 4	0.0445	0.1564	7,46	1.2

According to the obtained values of the distribution of sample characteristics X^F_i (jT), approximating polynomial for the functions due to the method of least squares is selected.

To calculate the m_i coefficients, the system of equations with four unknowns are used (eq. 5):

$$\begin{cases} 0{,}0084 = m_0 + 12m_1 + 12^2 m_2 + 12^3 m_3 \\ 0{,}0404 = m_0 + 24m_1 + 24^2 m_2 + 24^3 m \\ 0{,}0166 = m_0 + 36m_1 + 36^2 m_2 + 36^3 m \\ 0{,}0446 = m_0 + 48m_1 + 48^2 m_2 + 48^3 m \end{cases} \quad (5)$$

Solving this system of equations, the values of coefficients are obtained: m_0= - 0.1905; m_1= 0.027; m_2= 0.001; m_3= 0.00005.

In such a way the calibration interval of measuring potentiometer PCB-4P is obtained which is equal to 60 months.

4. Conclusion

The calculation provided in the paper confirms that calibration interval adjustment of measuring instruments is a worthy process and is economically profitable for the company as the calculated period of time between verifications exceeds 5 times the interval which is recommended by the manufacture.

Despite the economic benefits of the adjustment of the calibration interval there is some critical information that cannot be ignored. The critical approach refers to setting initial calibration intervals and such criteria are to be considered before the adjustment process starts: stability of the equipment; the complexity and critically level of the measurements to be made; the frequency of the use of the measurement and test; the environmental conditions (such as dust, vibration, and temperature) in which the equipment is operated; the risk of damage or misuse; the qualification of the equipment operators and technicians; the mode of the measurement: automated or manual one; the possession of the equipment: company-owned or employee-owned one; any customer contract obligations in regard to calibration intervals; any regulatory agency obligations regarding calibration intervals; the risks associated with the improper use of the calibrated equipment.

So the calibration interval is considered the most important parameter of the metrological service of measuring instruments directly affect the level of uniformity of measurements. The smaller the calibration interval is, the higher the level of measurement trueness is. On the other hand, the smaller the calibration interval is, the greater the financial costs of the calibration of measuring instruments are, as well as the production costs associated with the withdrawal of funds from the measurement site. Thus, determining optimal value for the calibration interval is found quite important.

References

[1] Aitchison J and Dunsmore I R 1975Statistical Prediction Analisys (*Cambrige Universisty Press*)
[2] Nunzi E *et al* 2005 Stochastic and reactive methods for the determination of optimal calibration intervals *IEEE Transactions on Instrumentation and Measurement* **54 (4)** 1565 – 1569
[3] Bobbio A *et al* 1997 Monitoring the calibration status of a measuring instrument by a stochastic model *IEEE Transactions on Instrumentation and Measurement* **46 (4)** 747 – 751
[4] Lin K *et al* 2005 A gray system modeling approach to the prediction of calibration intervals *IEEE Transactions on Instrumentation and Measurement* **54 (1)** 297–304
[5] Bailey A *et al* 2015 The stability and calibration of water vapor isotope ratio measurements during long-term deployments *Atmospheric Measurement Techniques* **8 (10)** 4521–4538
[6] Vidoni P 2009 A simple procedure for computing improved prediction intervals for autoregressive models *Journal of Time Series Analysis* **30 (6)** 577–590
[7] Bjørnstad J F 1996 On the Generalization of the Likelihood Function and the Likelihood Principle *Journal of the American Statistical Association* **91(434)** 791–806
[8] Galtseva O V *and* Bordunov S V 2015 *IOP Conference Series: Materials Science and Engineering* **81 (1)** 012062 DOI: 10.1088/1757-899X/81/1/012062
[9] ISO 10012:2003 2003 Measurement Management Systems *Requirements for Measurement Processes and Measuring Equipment*
[10] ANSI/NCSL Z540–1–1994: 1994 Calibration Laboratories and Measuring and Test Equipment *General Requirements*
[11] Gmurman E 2008 *Probability Theory and mathematical statistics* (Moscow: Higher Education)
[12] Kobzar A I 2006 Applied mathematical statistics *For engineers and researchers* (Moscow: Fizmatlit)
[13] Pritulov A M *et al* 2007 *Russian Physics Journal* **50 (2)** 187–192 DOI: 10.1007/s11182-007-0026-3
[14] Petrova A Y *et al* 2015 *Technical Physics* **60 (4)** 592–594
[15] Vasendina E A *et al* 2015 *IOP Conference Series: Materials Science and Engineering* **81 (1)** 012070 DOI:10.1088/1757-899X/81/1/012070

[16] Surzhikov A P *et al* 2009 *Nuclear Instruments and Methods in Physics Research, Section B* **267** **(7)** 1072–1076
[17] Surzhikov A P *et al* 1999 *Russian Physics Journal* **42** **(5)** 490–495
[18] Borikov V N *et al* 2016 *Journal of Physics: Conference Series* **671** **(1)** 012014. DOI: 10.1088/1742-6596/671/1/012014

AUTHOR INDEX

Institute of Physics
Dirac House, Temple Back
Bristol BS1 6BE UK

ISSN: 1757-8981
ISBN 978-1-5108-2535-2

CURRAN ASSOCIATES INC.
proceedings.com

Workshop on Language Resources, Technology and Services in the Sharing Paradigm 2011

Held at the 5th International Joint Conference on Natural Language Processing 2011 (IJCNLP 2011)

Chiang Mai, Thailand
12 November 2011